Security Practices Primer for Water Utilities

The mission of the Awwa Research Foundation (AwwaRF) is to advance the science of water to improve the quality of life. Funded primarily through annual subscription payments from over 1,000 utilities, consulting firms, and manufacturers in North America and abroad, AwwaRF sponsors research on all aspects of drinking water, including supply and resources, treatment, monitoring and analysis, distribution, management, and health effects.

From its headquarters in Denver, Colorado, the AwwaRF staff directs and supports the efforts of over 700 volunteers, who are the heart of the research program. These volunteers, serving on various boards and committees, use their expertise to select and monitor research studies to benefit the entire drinking water community.

Research findings are disseminated through a number of technology transfer activities, including research reports, conferences, videotape summaries, and periodicals.

Security Practices Primer for Water Utilities

Prepared by:
Michael Baker Corporation
3601 Eisenhower Avenue, Suite 600, Alexandria, VA 22304

Jointly sponsored by:
Awwa Research Foundation
6666 West Quincy Avenue, Denver, CO 80235-3098
and

U.S. Environmental Protection Agency
Washington D.C.

Published by:

Awwa Research Foundation
Advancing the Science of Water®

IWA Publishing

DISCLAIMER

This study was funded by the Awwa Research Foundation (AwwaRF). AwwaRF assumes no responsibility for the content of the research study reported in this publication or for the opinions or statements of fact expressed in the report. The mention of trade names for commercial products does not represent or imply the approval or endorsement of AwwaRF. This report is presented solely for informational purposes.

Published by IWA Publishing, Alliance House, 12 Caxton Street, London SW1H 0QS, UK

Telephone: +44 (0) 20 7654 5500; Fax: +44 (0) 20 7654 5555; Email: publications@iwap.co.uk;
Web: **www.iwapublishing.com**

AwwaRF report number 90984F. Originally published by AwwaRF for its subscribers in 2004.
IWA Publishing version published 2005.

ISBN 978-1-84339-888-2

CONTENTS

* Appendix A ends on p. 66 and the References begin on p. 447. There are no pages missing. Appendix B on CD begins on p. 67 and ends on p. 446.

LIST OF FIGURES

FOREWORD

The Awwa Research Foundation is a nonprofit corporation that is dedicated to the implementation of a research effort to help utilities respond to regulatory requirements and traditional high-priority concerns of the industry. The research agenda is developed through a process of consultation with subscribers and drinking water professionals. Under the umbrella of a Strategic Research Plan, the Research Advisory Council prioritizes the suggested projects based upon current and future needs, applicability, and past work; the recommendations are forwarded to the Board of Trustees for final selection. The foundation also sponsors research projects through the unsolicited proposal process; the Collaborative Research, Research Applications, and Tailored Collaboration programs; and various joint research efforts with organizations such as the U.S. Environmental Protection Agency, the U.S. Bureau of Reclamation, and the Association of California Water Agencies.

This publication is a result of one of these sponsored studies, and it is hoped that its findings will be applied in communities throughout the world. The following report serves not only as a means of communicating the results of the water industry's centralized research program but also as a tool to enlist the further support of the nonmember utilities and individuals.

Projects are managed closely from their inception to the final report by the foundation's staff and large cadre of volunteers who willingly contribute their time and expertise. The foundation serves a planning and management function and awards contracts to other institutions such as water utilities, universities, and engineering firms. The funding for this research effort comes primarily from the Subscription Program, through which water utilities subscribe to the research program and make an annual payment proportionate to the volume of water they deliver and consultants and manufacturers subscribe based on their annual billings. The program offers a cost-effective and fair method for funding research in the public interest.

A broad spectrum of water supply issues is addressed by the foundation's research agenda: resources, treatment and operations, distribution and storage, water quality and analysis, toxicology, economics, and management. The ultimate purpose of the coordinated effort is to assist water suppliers to provide the highest possible quality of water economically and reliably. The true benefits are realized when the results are implemented at the utility level. The foundation's trustees are pleased to offer this publication as a contribution toward that end.

Water utilities are increasingly turning to automation of their treatment facilities to help control costs and maintain or improve the quality and consistency of their treated water. An investment in automation requires a commitment to maintenance and staff training, and it is generally agreed that there is an appropriate level of automation depending on the plant's circumstances. To guide utility managers, design engineers, and regulatory officials in selecting the right levels of automation and staffing, this report establishes a decision-making framework and examines various factors that contribute to successful plant automation projects.

Edmund G. Archuleta, P.E.
Chair, Board of Trustees
Awwa Research Foundation

James F. Manwaring, P.E.
Executive Director
Awwa Research Foundation

ACKNOWLEDGEMENTS

The authors of this report are indebted to the following water utilities and individuals for their cooperation and assistance in the project:

The individuals who participated in the Expert Workshop were Steve Schmitt, American Water; Kim Dyches, Utah Division of Drinking Water; Brad Stewart, Salt Lake City Department of Public Utilities; Marcis Kempe, Massachusetts Water Resources Authority; Randy Hawley, Portland Water; Mike Gritzuk and Madeline Goddard, Phoenix Water; Bruce Johnson, Tucson Water; Amy Purvis, Milwaukee Water Works; Greg Moser, Jefferson County (CO) Office of Emergency Management; Jan DeBoer, Colorado Department of Public Health and Environment; Alan Roberson, AWWA; Steve Dennis, Alameda County Water District; and Gay Porter DeNileon, AWWA. Many of these individuals also provided outstanding examples of security practices.

The Project Advisory Committee Members were Grace Robiou, Environmental Protection Agency; Scott Kelly, Jacksonville Electric Authority; and Carrie Lewis, Milwaukee Water Works. Ms. Robiou and Mr. Kelly also participated in the Expert Workshop.

Barry Schwartz of Sandia National Laboratories provided a presentation at the Expert Workshop, as well as information on the Lab's intensive security program. Robert Faulhaber of Access Network Solutions also provided a presentation on cyber security at the Workshop.

East Bay Municipal Utility District provided the expertise of Ray Riordan to serve as Utility Advisor to the project. Mr. Riordan provided many useful documents, as well as valuable counsel.

We also thank the many individuals and utilities that contributed information and examples of security practices to the project and who completed the project questionnaire.

EXECUTIVE SUMMARY

Among America's critical infrastructures, none is more fundamental to the health and welfare of our citizens than the systems used to collect, treat, and distribute potable water. Security is by no means a new concept for water utilities; however, until now water utilities typically have developed security practices with little guidance from the industry. Recognizing this need, the American Water Works Association Research Foundation (AwwaRF) directed Michael Baker Corporation to develop a primer of security-related practices applicable to the water industry. This project represents an initial assessment of water security practices at water utilities in order to address immediate security concerns. The product of this project should provide useful information to utility managers. The information will facilitate informed decision-making concerning what policies or procedures they may want to adopt when addressing security concerns.

Extensive research was conducted on a wide range of security measures related to the protection of water-related organizations and infrastructure. Research consisted of a review of both published literature and electronic information available on the Internet. In addition, the Project Team reached out to organizations and stakeholders who may have information relevant to security and anti-terrorism preparedness. A questionnaire was designed to solicit specific security information that would be useful to the water industry. This tool was distributed to 82 water utilities and 27 representatives from other industries with proven security practices. The findings from the literature review and outreach were then reviewed in a workshop involving select managers from public and private water utilities, the Project Advisory Committee (PAC), State representatives, a representative from Sandia National Laboratories, and contractor and AwwaRF staff.

FINDINGS

This report is among the first guidance documents developed in support of security practice programs. It is evident that this is merely the beginning of addressing such programs, and that additional guidance is currently under development, which will contribute to a larger body of information available to the utilities. The guidance contained in this report addresses the following security practices:

- *Overall utility security policy*, forming the foundation of all security actions and establishing organizational and individual responsibilities for security.
- *Employee background checks* to protect the utility against employment of potential wrongdoers and liability for negligent hiring.
- *Employee and non-employee access and badging* to control who gains access to utility facilities and ensuring that only authorized individuals can access sensitive areas.
- *Training for security* to acquaint employees with sound security practices and the utility's security plan and test security plans.
- *Public tours*, providing guidelines for controlling the access of the public to facilities, particularly in times of heightened threat.
- *Alarm and CCTV monitoring protocols*, establishing an appropriate response when intruders are detected.

- *Threat levels*, presenting a staged set of security measures taken in reaction to increasing threat levels.
- *Chemical delivery and testing* to reduce the opportunity for the use of chemicals used in the treatment process as a potential pathway for contamination.
- *Package/mail delivery* to identify and control suspicious packages.
- *Guard service*, defining protocols for deployment of guards and use of deadly force to protect utility facilities.
- *Construction site security* for enhancing security at construction sites, which are often poorly secured.
- *Neighborhood watch*, employing citizens residing near utility facilities to serve as additional "eyes and ears" for the utility.
- *Public and media relations*, providing guidance on communications with the public and the media before, during, and after an incident.
- *Mutual-aid and regional cooperation*, planning for and employing the resources of multiple utilities in an area in the event of an incident.
- *Incident reporting, notification, and response*, presenting procedures and checklists for reporting incidents to the appropriate utility staff and external agencies and mounting an effective response.
- *Coordination with health/medical and pharmacies*, establishing mechanisms for the medial community and pharmacies to report suspicious outbreaks of illness.
- *Employee safety* to protect employees in an incident.
- *Response to actual and threatened contamination*, defining methodologies for validating threats, determining the hazard involved, and responding with appropriate resources if contamination with chemical, biological, or radiological agents is suspected.
- *FOIA and sunshine laws*, discussing the status and impact of such laws, which may affect the capacity of a utility to protect sensitive information.
- *Secure communications & Cyber security* for control of sensitive real time information during an incident, as well as protecting SCADA and business IT networks from intrusion.
- *Protection of sensitive information*, such as facility plans, system diagrams, GIS data, etc. from disclosure to individuals with disreputable intentions.

For each security practice, the report defines the purpose, general characteristics of a sound practice, implementation impediments, and legal considerations. Appendix B includes examples of policies and procedures.

While many utilities provided security information that was utilized in the development of this report, it is evident that security practices are evolving. Information that is currently under development will need to be collected and analyzed to provide the industry with a comprehensive security practice program. Based upon newly proposed rules and regulations (i.e., Freedom of Information Act and Sunshine Laws, the June 2002 Bioterrorism Bill, and Department of Homeland Security Procedures for Handling Criteria Infrastructure Information 68CFR18523), additional guidance will be forthcoming and should be included in the development of procedures in support of security programs.

RECOMMENDATIONS FOR FURTHER RESEARCH

As utilities put into practice new security practices and gain experience with implementation successes and difficulties, the state of knowledge will evolve rapidly. Therefore, future updates of this report should be anticipated. Future guidance documents should include an examination of the newly proposed rules and regulations, guidance and supportive materials currently under development by Federal agencies. Based on the work performed in this project, it appears that the following security policies and procedures should be further studied and documented:

- Employee background checks practices
- Employee security training
- Employee safety procedures in the event of a security incident
- Package/mail delivery screening
- Coordination with the health and medical community and pharmacies
- Neighborhood watch programs
- Information and communications security

Successful implementation the security practices within water utilities requires consistent documentation of policies and procedures, on-going training, and periodic review and updates. This document is intended to support water utilities as they develop, implement, and review their own unique security practices by providing basic guidance and an overview of how specific security issues are being addressed within the industry.

CHAPTER 1: INTRODUCTION

1.1 PROJECT OBJECTIVES

Among America's critical infrastructures, none is more fundamental to the health and welfare of our citizens than the systems used to collect, treat, and distribute potable water. The open nature of our society and its infrastructure has led to a fertile ground for terrorists and other criminal elements. Criminal acts directed against water systems can take several forms, from large-scale introduction of toxins that can sicken or kill thousands to small-scale vandalism that interrupts the flow of water to a service area.

Prior to the tragic events of September 11, 2001, most water utilities concentrated their physical and operational security considerations on preventing minor theft and vandalism. On September 12, 2001, all critical infrastructure sectors began viewing the security of their operations and facilities from a new perspective. While security is a widely understood concept, the discipline of providing security for critical infrastructures continues to develop. Presidential Decision Directive 63, of May 1998, anticipated a "public-private partnership" as a necessary requirement for protecting critical infrastructures. The details and workings of such a partnership are only being defined at this time. While much has been done since September 11, 2001 to raise awareness and understanding of security concerns, limited information has been made available on policies and procedures a utility might consider adopting to strengthen and improve water security programs.

The objective of this project is to provide utility managers and the water community with a comprehensive, easy-to-use primer on security-related *policies and procedures.* It must be emphasized that the study did not examine industry practices for physical security measures, such as alarms, cameras, fencing, etc. Although the project concentrated on water-related organizations, a number of other industries with similar facility and organizational level concerns were also examined to identify current and proposed security/anti-terrorism practices, methods, and know-how. The study focused on how these organizations incorporate security procedures into routine operational areas such as human resources, information dissemination, emergency response, media relations, and community coordination.

The product of this project should provide useful information to utility managers, facilitating informed decision-making concerning what policies or elements of these policies they may want to adopt when addressing security concerns. The practices described in this report are directed mainly toward public water systems serving more than 3,300 customers, although some of the basic policies and procedures could also apply to smaller utilities. In addition, the report provides a range of potential policies and procedures that utilities might consider adopting. However, not every practice will be appropriate to individual utilities. Each utility should evaluate whether a given practice is appropriate to them based on their specific situation, particularly the size of the utility and its identified threats.

Security in the water industry is continually evolving. This project represents an initial assessment of water security practices at water utilities in order to address immediate security concerns. The *Security Practices Primer for Water Utilities* is intended to be one of many tools utilities use to determine their own policies and procedures. As utilities implement new security practices, gain experience, and recognize successes and difficulties, the state of knowledge will evolve rapidly. As the body of knowledge grows, additional resources will become available. Therefore, future updates of this report should be anticipated.

1.2 ORGANIZATION OF THE REPORT

This report includes five parts:

- *Executive Summary*, provides an overview of the project and its findings.
- Chapter 1, *Introduction*, describes the overall purpose of the project and how it relates to other water security work. This section also outlines project activities and documents the outreach and research performed by the contractor to identify and collect candidate security practices for water utilities.
- Chapter 3, *Security Practices for Water Utilities – Results and Analyses*, provides a discussion of key security practices, including examples of policies and procedures implemented by water utilities and other organizations.
- Chapter 4, *Conclusions,* presents a summary of the project findings and recommendations for further study and research.
- *Appendices* provide a variety of background documents.

1.3 STUDY DESIGN AND METHODOLOGY

The core of the project was the gathering of information and materials that address security and anti-terrorism measures as they relate to safeguarding water-related organizations and infrastructure. This information collection was accomplished through two interrelated activities, a comprehensive literature search and an outreach to organizations that might have such information.

The first task in the process was a comprehensive search of available literature. Because there are potentially tens of thousands of sources, a review of all these sources was beyond the scope of this study. Therefore a targeted literature search, focusing on current and relevant sources of information, was used for this project. The emphasis, of course, was on water utility security, although other critical infrastructures were also examined. These publications provided the foundation for this research and were reviewed for applicable security practices in the key areas of policy, human resources, physical and operational security, coordination, technical and contamination response, and information and communications security. Ultimately, over 100 publications and periodical articles were identified and reviewed for this project.

Another source of pertinent information was the Worldwide Web. To focus the exploration, search terms common to the industry were used to initially locate information. In addition to keyword searches, specific websites of government agencies and water publications, water-related organizations, and utilities were also reviewed. Over 50 websites were reviewed. Interestingly, much of the information focused on the need for improved security measures. Several sources listed specific measures or recommendations.

The second component of information collection was an outreach to a wide variety of relevant organizations and stakeholders that may have information relevant to security and anti-terrorism preparedness. This effort was intended to uncover information on policies and practices not available from published material. During this process, as one source identified another potential source of information, that source was contacted. Organizations contacted included water industry associations; individual water utilities; law enforcement agencies; State emergency management, drinking water, and public health agencies; private sector corporate security representatives; and Federal agencies involved in infrastructure protection, including the Environmental Protection Agency and Department of Homeland Security.

A questionnaire was developed addressing specific elements of information most useful to the water industry. Questions were designed to elicit the best practices from a variety of industries and organizations. Input on the design of the questionnaire was solicited from the AwwaRF Project Manager and the PAC. AwwaRF sent letters of introduction to selected organizations to facilitate responses. Electronic questionnaire forms were e-mailed to 82 water utilities and 27 representatives of other industries with proven security practices. Over 50 candidate security policies and procedures were submitted by the utilities.

The findings from the literature review, benchmarking analysis, and outreach formed the basis for a facilitated workshop involving selected system managers from public and private water utilities, the PAC, State representatives, a representative from Sandia National Laboratories, and AwwaRF staff. The workshop presented findings from the literature review, benchmarking analysis, and outreach and provided an exchange of technical expertise from leading security experts. Participants were given the opportunity to validate the most innovative and practical security and anti-terrorism procedures. All pertinent comments and recommendations from the participants were recorded and reviewed by the Project Team and are presented in this report.

The literature review, benchmarking analysis, outreach, and participant discussions from the workshop formed the basis for the *Security Practices Primer for Water Utilities.* This document presents the culmination of all findings from this project, including a separate chapter on recommendations to utilities. These recommendations were based on how practical and applicable these findings are for incorporation into water system security procedures.

CHAPTER 2: SECURITY PRACTICES FOR WATER UTILITIES

2.1 OVERVIEW OF SECURITY PRACTICES

An integrated security system has three elements: physical security, operational security, and IT network security. These three "pillars" of security and the key components of each are shown below. The objective of security is to deter a criminal's intent by slowing or preventing the penetration, detecting it early enough to prevent the success of the attack, or mitigating the effect of the attack if it cannot be prevented. The three pillars are interrelated. This project concentrated upon the middle pillar, operational security, including policies and procedures.

Figure 2.1: Integrated Security System

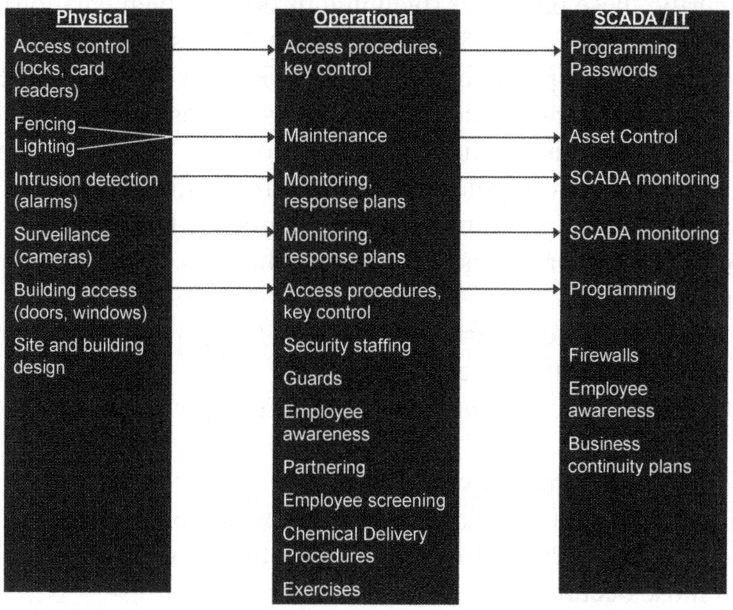

Underlying all security is the culture of the organization and its employees. If senior management does not provide resolute and unambiguous support for security, along with the necessary resources, employees will not take security seriously. Likewise, employees can defeat the most sophisticated security system with carelessness or apathy - leaving doors unlocked, sharing access badges, and not being sensitive to confidential information. Employee and management training and awareness are critical elements of any security structure.

In developing an integrated security program, the principle of trust is central. Employees will participate in security solutions that they understand and support and will then accomplish their mission securely. Employees must have a role in the development of security policies and practices at the utility. However, the success of solutions must be verified in an appropriate manner.

In developing an effective security program, the following should be considered:

- Develop and follow a set of guiding principles for the program development and implementation.
- Have clear objectives for the Security Program and communicate that to management and employees.
- Define clear performance standards (how good is good enough?).
- Metrics, or measures of performance, should allow management to know how they are doing against defined performance standards.
- The Security Office and employees (the persons who implement) partner to develop and improve the system.
- When considering whether or not to endorse proposals for changing business processes to accommodate security, these processes (planning through implementation) should:
    - Be the minimum set necessary to enable our employees to contribute to the utility's water supply mission, to operate in accordance with its corporate values, to be legally and contractually compliant, and to take advantage of best business practices.
    - Specify what is to be achieved, and, wherever possible, how it is to be achieved will be left up to operational elements within the utility to decide. (Uniform "how" specifications will be imposed only where not to do so would result in unacceptably high corporate risk or where there is a high enterprise cost advantage for doing so.)
    - Be based on the presumption of trustworthiness. To support this presumption:
        - Communicate expectations clearly;
        - Provide training, tools, and support to enable employees to meet expectations; and
        - Hold employees accountable consistently.
    - Be based on an evaluation of risks - identifying what could hinder or prevent the utility from meeting its planned objectives and analyzing the consequences and probability of those occurrences.
- Periodically review policies and processes to determine if they have met the stated objectives and take action if they have not.
- The operational elements of the utility "own" security implementation – not the security organization. NOTE: Recognize that the security organization may wear "two hats," but when they are performing security tasks, they are in a "line capacity". The Security organization has responsibility for the security "program."
- Involve legal and diversity/human rights components of the utility early in the process to ensure compliance with applicable laws and regulations and avoid actions that may offend certain groups of employees or customers.
- Recognize that zero risk is impossible, so tailor security procedures to the likely threat.

2.2 OVERALL SECURITY POLICY

Objective

The foundation of an effective security program at a water utility is an overarching statement of policy in support of security adopted by the highest authority in the utility, such as the Board of Directors. Policy statements supporting workplace and customer safety have been standard utility practice for years. However, many utilities had not considered security as critically important until recently, and so had not adopted an over-arching security policy. Such a policy should express in clear and unequivocal terms that security is essential to the safety of the water supply provided to customers and is a primary function of the utility. The policy fosters the establishment of a culture of security among utility management and employees. Further, a security policy helps reassure the public and government officials that their water supply is safe.

Summary of Elements

Some elements to consider in developing a security policy are:

- Support for security should be stated unambiguously.
- The policy should be adopted by the highest authority in the utility and implemented by directives from senior management.
- Objectives of the security program should be stated.
- The senior managers themselves should assume ultimate responsibility for security.
- Responsibility for development and management of the security program at the utility should be specifically assigned to one organization or individual.
- Adequate resources, both staff and financial, should be provided to the security program.
- All employees should be held responsible for following security procedures in their duties. Security should be an element of each employee's performance evaluation. Violations of security should be grounds for disciplinary action.
- The policy should be coordinated with higher authority within the municipal government, if appropriate.
- The policy should be coordinated with neighboring water utilities, particularly when such utilities are interconnected, and with regulatory, law enforcement, and emergency management agencies.
- The policy should be reviewed by the utility's legal, human resources, and diversity organizations as it is being developed.

In developing the policy, employees must be held accountable for security. For example, Sandia National Laboratories requires that all employees:

- Work safely and securely.
- Know how their actions impact the safety and security of others.
- Protect each other, the public, and the environment.
- Protect laboratory assets (information and property).

- Obey signs, barricades, labels, and tags.
- Ask questions when in doubt and communicate any security concerns.
- Refrain from participating in activities they believe are unsafe or may compromise security.
- Question persons whose authorization or access privileges are not known or who don't appear to be working securely.
- Complete security training required for the job and work area.

Examples of security policies that have been successfully implemented are included as Exhibits 1 and 2.

Implementation Impediments

If a utility is unionized, the policy may have to be negotiated with the unions prior to adoption, particularly the responsibilities of employees and the grounds for disciplinary action. In addition, security actions may be restricted by public access requirements. For example, at one major water wholesaler, the entire grounds of the headquarters campus is open to the public as a park by agreement with the city parks department.

Legal Considerations

As municipal or government agencies, some water utilities may be required to conform to the security requirements of their higher authority. Federal, State, and local laws and regulations may also apply. In all cases, a legal authority should review the security policy.

Examples of Utility Security Practices and Other Background Materials

These documents are located in section 2.2 of Appendix B.

A. Exhibit 1 is the Security Policy implemented by a major California water utility.

B. Exhibit 2 is a letter sent to all employees of the Phoenix Water Services Department describing a zero tolerance policy for security breaches.

2.3 HUMAN RESOURCES

2.3.1 Employee Background Checks

Objective

The security threat posed by employees can be one of the most difficult to deter. Moreover, since employees often have access to the most sensitive information and assets of the water utility, it is important to identify this potential threat at the earliest possible moment. The good news is that the security threat posed by prospective employees may be detected and prevented in many cases.

Detection and prevention of the security threat posed by an employee begins before the employee is hired. Screening through appropriate background checks will assist the employer with early identification of individuals who have a criminal past, or a history of other behavior that could pose a security risk to the water utility. The objective of employee background checks is to provide the employer with (1) identification of employee candidates who could pose a security risk, and (2) an opportunity to avoid the added security risk by not hiring a particular employee candidate. Ideally, background checks should also be performed periodically on existing employees to identify problems and behaviors that could compromise their integrity.

Summary of Elements

In order to identify and avoid the hiring of a person who could pose a security threat to the employer, an employee background check:

- Must be conducted before the prospective employee becomes an employee,
- Must be appropriate and thorough for the particular position of employment,
- Must be reviewed and consistently applied to all prospective employees, and
- Must be lawfully conducted.

Background checks are most appropriately conducted before a prospective employee is hired. In some cases, it may be appropriate to tailor the background check to the sensitivity of the position for which the individual is applying. Background checks should also be performed on existing employees. The events of September 11, 2001 demonstrate that potential terrorists are willing to spend considerable time preparing for an attack. This could include seeking employment at potential targets, such as utilities. In addition, employees may be developing financial or behavioral problems or a criminal record that could compromise their integrity. Although not perfect, a background check could identify these issues before they become a problem for the utility. Utilities should consider conducting a one-time background check on all existing employees just to ensure that these employees haven't assumed a false identity. Again, checks should be tailored to an individual's duties, as is done in the Federal Government. Individuals in more unskilled positions, such as janitors and clerks, might have access to very sensitive areas, and thus require more scrutiny than higher-level personnel without access to such assets.

A survey published by the Society of Human Resource Management in 1998 gave the following results for resumes and job applications where falsification had been detected:

- Length of employment 53%
- Past salaries 51%
- Criminal records 45%
- Former job titles 44%
- Former employers 35%
- Driving records 33%
- Degrees 30%
- Credit 24%
- Schools attended 24%
- Social security number 14%

Department of Labor statistics show, surprisingly, that 73% of all current drug users age 18 or older are employed, with over 80% employed full-time. Businesses without a drug-screening program are naturally more likely to hire such individuals and to suffer the hidden costs associated with them. A U.S. Postal Service study indicated that employees who tested positive on their pre-employment drug test were 77% more likely to be discharged within the first three years of employment and were absent from work 66% more often than those that tested negative.

Employers can successfully perform background checks in-house. However, there are many companies that specialize in conducting pre-employment background checks and they may offer valuable assistance in selecting the appropriate level of background check for the particular employee position to be filled. Whether background checks are performed in-house or by a company specializing in such matters, the employer must be careful that the background checks are conducted in compliance with applicable state and federal laws. Turnaround times for these checks range from the immediate (with online access to credit and other up-to-date databases) to generally two or three working days, depending on the type and extent of search desired. Fees are surprisingly low and can range from $3, up to $50-80 depending on several factors: the package of searches required for a particular position, the number of references contacted and the number of counties searched for criminal activity.

Background checks can vary in level of detail depending on the security needs of the employer and the particular employee position to be filled. The following is a list of elements that could be included in a background check:

- Driving records
- Vehicle registration
- Credit records
- Criminal records
- Social security number
- Education records
- Court records
- Worker's compensation
- Employment verification
- Character references

- Neighbor interviews
- Medical records
- Property ownership
- Bankruptcy
- Military service records
- State licensing records

The extent of the check can be personalized to the employee's duties. Background checks must be carefully and consistently reviewed and applied. It is imperative to the success of the program that employees at all levels of the organization be treated fairly. Advice from legal and diversity/human resources staff should be considered in the design of the program.

One issue that must be considered in determining whether or not to implement employee screening is the obligation of the utility to its customers and stakeholders. Courts have held employers liable for negligent hiring, when they fail to ensure that their employees are trustworthy. For example, an Oakland civil jury awarded over $11 million to a woman's husband after she was murdered by a man who came to clean their carpets. Under the legal doctrine of "negligent hiring" the carpet care firm was ordered to pay over $9 million of the award despite their defense that the perpetrator was an independent contractor and not an employee. A delivery firm recently made headlines in Los Angeles when one of its home delivery drivers, two weeks on the job, allegedly committed sexual assault on a San Gabriel woman as well as attacking two other women. The firm had failed to check the criminal history of the employee who, unknown to them, was on parole for a violent conviction. In another recent instance, a home health care provider in Massachusetts unknowingly hired someone with a criminal record as a home attendant. That individual subsequently murdered a young paraplegic man placed under his care, as well as the patient's grandmother, apparently to cover up a theft. When the employee was later tried and convicted of the crime, the victim's family sued the company on the basis of negligent hiring and obtained damages of $26.5 million dollars. Not surprisingly, the firm is no longer in business. More information on negligent hiring appears below and in the examples.

Discussion of Implementation Impediments

The issue of background checks can be a very sensitive topic. There can be much concern over the confidentiality of information viewed during a background check. Specific concerns may arise over how the information will be used, how the information will be stored, who will have access to the information, and if it will become a permanent part of a person's employment record. Unions or individual employees may express objections to background checks. In some states, conducting background checks on existing employees may not be legally permissible without their prior consent.

Legal Considerations

All aspects of the employer-employee relationship require consideration of the laws and industry practices in order to limit exposure to claims or lawsuits by prospective employees, existing employees, and other third-parties. While there is no substitute for sound legal advice by an attorney competent in employer-employee relations, the following paragraphs provide an overview of the areas to be considered.

Background Checks and Negligent Hiring

Negligent hiring is a claim for damages resulting from an employer's alleged failure to investigate and discover a prospective employee's propensity for certain behavior likely to cause damage or injury to another if the person is hired by the employer. Generally, in states where the negligent hiring claim is allowed, an employer may be found liable for negligent hiring when an employee causes damage or injury to a coworker, customer, or member of the general public.

Generally, negligent hiring lawsuits are based on the failure of an employer to conduct appropriate background research that would have disclosed the employee's past misconduct. Therefore, the employer can be ruled negligent for placing a person with criminal or malevolent tendencies in a position of trust where he/she could pose a threat to others. This liability of the employer will exist whether or not the employee is acting within the scope of his/her job duties, and/or whether or not the harm is done to a coworker, customer, or member of the general public.

If a third-party screening company is hired to conduct background checks, then it is prudent for the employer and its attorney to analyze the documentation provided by the screening company to verify compliance with federal and state laws, such as written notice, authorization, disclosure, and adverse action procedures.

Use of Consumer Reports for Background Checks

With regard to prospective employees, certain rights and procedures have been established to protect against discrimination and misuse of information collected during the pre-employment process. The use of background checks raises the question, how to do we use background checks effectively and lawfully? The answer begins with the protections afforded by our consumer protection laws.

Consumer reporting is a highly regulated area with the primary requirements for appropriate disclosures and reviews of a person's history being codified in the Fair Credit Reporting Act (FCRA). The Federal Trade Commission enforces the FCRA, and it is designed to promote accuracy and ensure the privacy of the information used in consumer reports. The full text of the FCRA is available online at http://www.ftc.gov/os/statutes/fcra.htm. Helpful information for both employees and employers regarding the FCRA is available online at http://www.fair-credit-reporting-act.com. Employers should also check with their attorney for specific state laws that may affect proper conduct of background checks and handling of the information gathered by the process.

The FCRA provides the procedures to be followed with regard to "consumer reports." A consumer report contains information about a person's personal and credit characteristics, character, general reputation, and lifestyle. To be covered by the FCRA, the report must be prepared by a consumer-reporting agency (CRA). A CRA is a business that assembles such reports for other businesses.

Employers conducting pre-employment background checks and/or obtaining consumer reports during a person's employment may want to want to review a person's credit payment records, driving records, and/or criminal history. Employers may order investigative consumer reports - reports that include interviews with a person's friends, neighbors, and associates. All of these types of reports are "consumer reports" if they are obtained from a CRA.

In addition, a prospective employee may be asked to give references. A reference verified by the employer is not covered by the FCRA; a reference verified by an employment or reference-checking agency (or other CRA) is covered by the FCRA. Section 603(o) provides special procedures for reference checking; otherwise, checking references may constitute an investigative consumer report subject to additional FCRA requirements.

Employers may use consumer reports when hiring new employees and when evaluating employees for promotion, reassignment and/or retention **so long as** the employer complies with the FCRA. Sections 604, 606 and 615 of the FCRA describe the employer's responsibilities when using consumer reports for employment purposes. Two of the most important procedures that employers must follow, include:

Written Notice and Authorization. Before an employer orders a consumer report for employment purposes, the employer must give *written notice* to the person (in a document consisting solely of this notice) that a report may be obtained and used by the employer. The employer also must get the person's *written authorization* before requesting a CRA for the consumer report.

Adverse Action Procedures. If an employer relies on a consumer report for an "adverse action" (e.g. denying a job application, reassigning or terminating an employee, or denying a promotion), the employer must follow these two steps:

1. Before taking the adverse action, the employer must give the person a *pre-adverse action disclosure* that includes a copy of the individual's consumer report and a copy of "A Summary of Your Rights Under the Fair Credit Reporting Act" - a document prescribed by the Federal Trade Commission. The CRA that furnishes the person's consumer report should provide the summary of consumer rights.
2. After taking the adverse action, the employer must give the individual notice (orally, in writing, or electronically) that the action has been taken in an *adverse action notice*. This notice must include:
   a. The name, address, and telephone number of the CRA that supplied the consumer report;
   b. A statement that the CRA that supplied the report did not make the decision to take the adverse action and cannot give specific reasons for it; and
   c. A statement of the person's right to dispute the accuracy or completeness of any information the CRA furnished, and of the person's right to an additional free copy of the consumer report from the CRA upon request within 60 days.

Examples of Utility Security Practices and Other Background Materials

These documents are located in section 2.3.1 of Appendix B.

A. Exhibit 1 is a wide-ranging paper on negligent hiring prepared by Michael Baker Jr., Inc. for a water utility client.

B. Exhibit 2 is a very comprehensive Personnel Surety Program policy on hiring and background checks developed by a national investor-owned utility. This example includes a variety of policy documents, notifications to employees, and a bargaining unit agreement. Note that this program focuses on employees in sensitive positions only.

2.3.2 Employee and Non-Employee Access and Badging

Objective

The need to secure facilities requires accounting of all persons entering and exiting the site and select areas within the site. In addition, EPA requires that all persons on sites with hazardous chemicals be accounted for in case of an emergency. The objective of this section is to present effective badging policies for personnel and implementation of those policies.

Summary of Elements

Employee and contractor badging is a universally accepted way of providing a secure site. Advancements in technology have transformed simple photo ID badges into tools to electronically control access, to indicate chemical exposure or to monitor an individual's movement within a facility.

Even the most sophisticated badge system is ineffective if the policy for updating badges, issuing new badges, replacing lost badges or eliminating badges of terminated employees is not followed. The following bullet points outline recommended security practices derived from the exhibits included in section 2.3.2 of Appendix B, as well as from the discussions at the Expert Workshop:

- Badges must include current photo (updated annually), name and position/company.
- Badges must include easily identifiable logo.
- Badges must be displayed at all times and at critical entry points.
- Lost badge procedures must be documented.
- Visitor badge procedures and/or escorts with appropriate badges must be provided for visitors. The visitor badge should be distinctive in color or text to make it obvious that the individual is a visitor. Some utilities use a special disposable visitor badge on which all text disappears after 24 hours. In all cases, visitors should be advised of the areas where they are not permitted to enter.
- Badges coded or colored for restricted access areas recommended.
- Electronic access badges can be programmed for certain areas and time limits set on the electronic access.
- Guards/security personnel should maintain records of badge holders and copies of badges.
- A written policy on the issuance of badges to new personnel required.
- Badges should only be issued after background checks have been performed.

One critical element of access control is the collection of keys and badges when an employee leaves the utility or when a contractor or vendor no longer needs access. The utility should require that these items be turned in to the appropriate authority before the individual actually leaves the premises. If an individual fails to return a badge it should be deactivated immediately. If an employee will be separated from the utility as a result of disciplinary action or other special circumstance, the employee's supervisor should collect the keys and access badge immediately.

Discussion of Implementation Impediments

The cost of implementing and maintaining a badge system is an impediment. The level of sophistication of badges needs to be considered when implementing a badge system.

Legal Considerations

None noted.

Examples of Utility Security Practices and Other Background Materials

These documents are located in section 2.3.2 of Appendix B.

A. Exhibit 1 from Alameda County Water District provides general guidelines for the issuance of ID cards, requirements for displaying ID cards, procedures for lost or stolen cards and procedures for returning cards at the end of employment.

B. Exhibit 2 provides guidelines for the issuance of ID cards, requirements for displaying ID cards, procedures for lost or stolen cards and procedures for returning cards at the end of employment. In addition it outlines access controls applied to cards/employees.

C. Exhibit 3, from the same utility, defines the procedure for changing the access privileges of a badge or employee.

D. Exhibit 4 is a form from Phoenix used to change access privileges of a current ID badge holder.

E. Exhibit 5 is a badge application form from Phoenix.

2.3.3 Training For Security

Objective

An effective security training program should prepare employees, as well as emergency response agencies, to respond to security-related situations. Training should be based on the municipality's/utility's security and emergency response procedures.

Security training programs should include emergency scenarios (man-made and natural), day-to-day operations, procedure review, public relations, safety precautions, observation and awareness training, etc. The purpose of training is to provide information to employees related to operations and maintenance of utility. Reinforcement of procedures, methods, etc. will ensure that the system (including personnel) will run efficiently during emergency situations.

Summary of Elements

Employee security training based upon security considerations and the emergency response plan should be performed annually to adequately prepare personnel to respond rapidly to a variety of possible situations. Furthermore, personnel should be able to respond effectively to an emergency in the absence of direction from their supervisors. To supplement annual training, procedures should be reviewed periodically with key personnel.

To be effective, training must not merely filter down from management. Training must be interactive with personnel who are likely to be the first responders to an emergency incident. Depending on the resources available within a given utility, the utility may elect to offer security training in-house or out-source the training by bringing in external security experts.

Key elements for effective training:

- Training schedule – train frequently, more often for anticipated major special events (i.e. Olympics).
- Practice what is trained through exercises using a variety of scenarios. Debrief on the results of exercises and make changes to procedures when indicated by problems during the exercise, i.e., create an action plan.
- Make training mandatory. All pay raises, approved vacation, etc. should be tied to the completion of all required training with no exceptions, including for senior management.
- Train everyone. Ensure that lower tier employees witness management and supervisory personnel attending same training. No one is excluded.
- Use specialized training as reward system, for example, no promotions without position specific training.
- Involve trainees in the training through interactive presentations.
- Change presentations/trainers to keep trainees interested.
- Reward employees for attending training by providing snacks, lunches, etc.
- Train or at least touch on all operations and facilities of the utility to expose all employees to what others do within the utility. Develop more intensive facility-specific training for employee assigned to that facility.
- Specialty topics, such as emergency response or terrorist threats, should involve outside agencies, such as the police.

Utilities' training schedules and needs vary depending on the size of the utility, number of employees, population served, etc. Smaller utilities may need less training than larger utilities. Areas that are host to large events (i.e. international sporting events) will need to perform specialized training to incorporate public relations, traffic and crowd control, language/culture awareness, etc.

Discussion of Implementation Impediments

The largest problem with implementing training is meeting people's schedules, especially for a utility that operates 24/7. Training should be scheduled well in advance so employees can plan ahead and rearrange schedules as necessary. Several of the same training session could be scheduled at different times and days to ensure attendance by all required.

Hiring procedures/paperwork should communicate the expectation that all employees will attend training – anticipated training schedules, types, durations, etc. could be included in new hire packages and employee newsletters.

Legal Considerations

Due the sensitive nature of potential security or emergency response training topics, employees may be required to sign nondisclosure agreements or be subject to disciplinary action, including dismissal, for disclosure of information outside the utility. Some training topics may only be available for specific responsibilities or positions. As with new hires, background checks may be required prior to training.

Examples of Utility Security Practices and Other Background Materials

These documents are located in section 2.3.3 of Appendix B.

A. Exhibit 1 is an outline from a utility's training seminar in preparation for a major international sporting event:

2.4 PHYSICAL AND OPERATIONAL SECURITY

2.4.1 Public Tours

Objective

Prior to September 11, 2001, most utilities encouraged an "open" atmosphere with the public. In many cases, water utilities are a public utility, so felt an obligation to the public to allow their access to utility operations. Tours or field visits at utility facilities are conducted for many reasons:

a. Benefit the utility's interest in serving customers, such as educating legislators to gain public support on utility projects.
b. Provide knowledge to other water utilities on system operations.
c. Educate the general public, including schools, service clubs, and youth organizations, with the goal of demonstrating the safety of their water.

After September 11, some utilities cancelled all public tours. However, the fundamental value in providing the public with some access remains. With a requirement for heightened security at water utilities, there is a need to establish procedures for managing the process of requesting and providing tours of utility facilities to interested public groups, while maintaining security awareness and minimizing the potential of impact on utility operations. The objective of this section is to present effective public tour security policies and implementation.

Summary of Elements

Based on the experience at utilities surveyed, the following regulations could be imposed on tours:

- All facility tours should be arranged sufficiently in advance to allow security screening and pre-approval.
- Persons or groups requesting a tour should be provided with a tour request form, including tour guidelines. This form, including the names and affiliations of all tour participants and an itinerary with the dates, times, and site locations of the visit, should be submitted to the utility 2 weeks prior to the tour for approval.
- Upon approving a tour, the utility should issue the interested public group a letter on utility letterhead.
- A point-of-contact of the interested public group should be designated and responsible for working with utility staff to make arrangements and provide necessary information.
- Transport vehicles should gain entrance through a security checkpoint and be inspected.
- The point-of-contact of the public group should present the approval letter issued by the utility.
- Tour participants should pass through a security-scanning device (airport scanning device or wand).

- A picture ID should be required for every tour participant upon arrival to the facility. This ID should be verified against the participant list submitted to the utility. Acceptable identifications are a valid driver's license, passport, government ID, or student ID card. In the case of groups with young children who may not have a picture ID, the chaperones should provide an ID and verify the identities of the children.
- All tours should be conducted with an assigned utility escort. Tour participants should not be allowed to roam freely at any time.
- The number of tour participants should be limited to a maximum of 10 participants. In the case of larger group, additional escorts should be assigned or participants divided into smaller groups.
- Student groups of children under the age of 16 should have adult chaperones.
- A security and safety briefing should be provided to participants prior to the start of the tour.
- No detailed facility drawings should be distributed or made available to tour participants.
- No still or video cameras, tape recorders, or any other electronic recording devices should be allowed on the tour without prior approval by the utility.
- The following items should be strictly prohibited and not allowed into the water utility facilities: backpacks, book bags, satchels, large briefcases or bags, aerosol sprays, any pointed objects such as knitting needles, scissors, sharp metal nail files, sharp objects such as knives, razorblades, box cutters, any guns, ammunition, fireworks, electric stun guns, mace, or martial arts weapons/devices. The utility should reserve the right to prohibit any other personal items.

In response to the Homeland Security Threat Level Advisory System, water utilities could institute public tour guidance consistent with the following procedures:

- Level Green or Blue Security: Tours can continue. In addition to the normal requirements, tour participants must present a form of personal identification that the utility staff can document.
- Level Yellow Security: Tours limited to those that directly benefit the utility's ability to serve customers. In addition to Level Green or Blue requirements a utility escort to tour participant ratio of 1:5 should be met.
- Level Orange or Red Security: All tours cancelled.

Discussion of Implementation Impediments

The utility could anticipate some negative publicity if tours are curtailed, particularly from elected officials. Some groups may resist the imposition of more stringent requirements as an invasion of privacy. Against this, the utility should stress the critical nature of the public's water supply and the need to protect the supply in a dangerous world. The utility might want to consider mounting a media campaign before implementing tour security.

The cost of security-scanning devices is an impediment. The level of the sophistication of security scanning devices needs to be considered when implementing this system.

Legal Considerations

None, unless the municipality's laws or regulations or basic utility agreements require municipal agencies to conduct public tours.

Examples of Utility Security Practices and Other Background Materials

These documents are located in section 2.4.1 of Appendix B.

A. Exhibit 1 is the procedures used by Alameda County Water District, including critical facility tour request guidelines and a facility tour request and guideline information form.

B. Exhibit 2 is the Agency Tour Protocol used by Massachusetts Water Resources Agency.

C. Exhibit 3 is a facility tour policy from a major California water and wastewater utility.

2.4.2 Alarm and Closed Circuit Television Monitoring and Response

Objective

The purpose of this section is to establish a policy and procedure for security alarm and closed circuit television (CCTV) monitoring and response. This section does not attempt to present detailed technical information on security alarm and CCTV monitoring system design and architecture. Detailed technical information on alarm and CCTV system design and architecture can be obtained from numerous security system designers/vendors. More importantly, the included security practices focus on the appropriate response to threats triggered by alarms or CCTV monitoring to ensure the protection of employees, facilities, and finished drinking water.

Security alarms and CCTV can be an effective tool for detecting potential and actual intrusions if operated correctly. An alarm or detection of a perceived threat has served no purpose if an appropriate response is not in place and executed in time to deter the criminal actions of an intruder or mitigate the effects of the attack. For instance, if an alarm at a remote site is activated and a timely response is not initiated, the potential threat has not been averted. There will typically also need to be some delay features built into the security system to slow down the perpetrator. Without the added security elements of delay and response, the alarm and CCTV detection/assessment system has not served its full purpose.

Summary of Elements

Alarm and CCTV monitoring and response policy should be directed toward:

- Employee Safety
- Detailed Response Protocols
- Clearly Defined Responsibilities

The most important aspect of the alarm and CCTV monitoring and response policy and procedure is employee safety. In regard to employee safety, the procedure may involve the following guidelines:

- The employee will attempt to assess the cause of the alarm from the safety of the interior of the facility.
- Unless properly authorized as part of a response force, the employee will not approach a suspicious vehicle or person.
- The employee will note descriptions of the person(s), vehicle, license plate numbers or any other identifying information from a safe and secure location.
- The employee will remain in the facility until the alarm has been assessed and the facility is determined to be secure by the security guard/response force.
- All communications between the employee and the security guard/response force will be via telephone, rather than radio, which can be monitored.
- If the alarm cannot be verified as non-criminal but the site has been secured, the employee may request a security guard/response force accompany him/her during subsequent duty rounds.
- Once confident in their personal safety, the employee will return to normal duty operations.
- If the employee suspects that the source of the alarm is criminal in nature and has direct concern for their own personal safety they will dial their local emergency response number, such as 911 or the utility's own armed response force, and request immediate emergency response from local law enforcement.

To ensure that security alarms and CCTV systems are able to provide for protection of employees and assets, the following protocols should be conducted on a regular basis, such as at the end of a normal workday or at shift changes:

- A walking security check and lock-down of facilities.
- Verification that all exterior doors are closed and locked.
- Verification that all alarms are set and cameras are operational.
- Verification that lighting is functioning properly.

An effective policy and procedure will incorporate detailed response protocols. It may be beneficial for the organization to develop different classes of incidents ranging from suspicious activity to confirmed criminal activity and develop detailed response protocols for each class. These protocols may include the following components:

- If the employee cannot determine the source of the alarm, he or she will contact the appropriate authority (i.e. security guard, police officer, etc…).
- The employee will include a description of the incident in the operation logbook.
- If a security guard or local law enforcement officer has been called, the employee will immediately notify his or her facility supervisor.
- The facility supervisor will take the appropriate steps to notify any other necessary staff and/or agencies.

- If the facility supervisor deems it necessary, the facility will be locked down.
- A potential chain of command notification for a municipality must be established. As an example, the facility supervisor will notify the superintendent, who will notify the general manager, who will notify the Mayor.

Finally, an effective policy and procedure will have clearly defined responsibilities. The clearly defined responsibilities may include the following:

- Employee
  - Report any suspicious activity
  - Note observations of suspicious activity
  - Notify security guard / response force
  - Notify facility supervisor
- Security Guard
  - Assist employee
  - Notify police
  - Notify plant staff
  - Notify facility supervisor
  - Lock down plant when directed by the facility supervisor
- Facility Supervisor
  - Respond to the scene immediately to supervise the incident response
  - Asses whether to lock down the facility
  - Asses whether to sample the clear well or other areas where finished water is exposed
  - Assess whether water distribution from the facility should be suspended
  - Assess whether additional personnel are required and contact them
  - Notify the superintendent
- Superintendent
  - Authorize the plant or water system shutdown and provide for alternate delivery or supplies
  - Notify the Health Department
- General Manager
  - Change operations as necessary
  - Issue a press release
  - Notify the Mayor

In addition, it may be appropriate for an organization to develop separate alarm and CCTV monitoring and response plans for "Normal Business Hours" and "After-Hours" if the response protocol varies between the two shifts.

Another issue is central vs. local monitoring of alarms and cameras. Most security experts suggest that a system should have both capabilities. At staffed facilities, the employees on site certainly need to be aware of potential or actual intrusions that might affect their safety and the operations at their facility. They will frequently be the first to identify and respond to an intrusion or the first to notify the response force. However, at many water system facilities, such as water treatment plants, only one employee may be on duty at any given time. These employees have other duties that may make it impossible for them to vigilantly monitor security or CCTV alarms. Further, the employee might be a victim of the intruder. Therefore, there should be a capability for monitoring of the site remotely, either from a central control room or another facility. Unstaffed facilities must obviously be monitored remotely. Here again, monitoring from two or more sites is desirable to provide a backup.

Discussion of Implementation Impediments

After a utility decides to implement the use of security alarms or CCTV, they must decide if the response to alarms or potential threats will be performed by internal employees or external organizations (i.e. private security or alarm firm). There are pros and cons for the use of both and a decision must be made by the individual organizations based on factors such as employee attitudes, geographic extent, and budgetary constraints.

For instance, if internal employees are used to respond to alarms they are generally familiar with the site, equipment, and process and can easily identify irregularities. However, the employees must receive training and frequent refresher courses on the proper response procedures. Several organizations have encountered resistance from unions or employees asked to respond to alarms.

If utility decides to use external organizations, such as a private alarm service, to respond to alarms, they should specify that responders be professionally trained to confront potential threats. Maximum response time should be specified. It should be noted that external organizations are generally less familiar with the site, equipment, and process and cannot easily identify irregularities; so contracted responders should receive familiarization training from the utility. In addition, responders from external organizations may have a delayed response time, unless they are contracted to be on-site.

Legal Considerations

As mentioned previously, some employees may belong to a union that does not allow members to be used for response to an alarm or a threat. Special legal consideration should be taken under these circumstances.

Organizations should also be aware of liability issues that may arise by using fake security cameras or signs making a statement such as "Site is monitored 24 hours a day" when in actuality it is not. The use of these measures may install a false sense of security in individuals. If an incident then occurs, the organization may be held liable.

Examples of Utility Security Practices and Other Background Materials

These documents are located in section 2.4.2 of Appendix B.

A. Exhibit 1 is a set of "Emergency Response Procedures" from a water utility in Utah.

B. Exhibit 2 is an "After-Hours Facility Alarm Activation/Response," from Alameda County Water District, which describes procedures to be followed at staffed facilities at the conclusion of the normal business day and during off hours.

C. Exhibit 3 is a "Procedure for Reporting Suspicious Incidents" from the City of Phoenix that establishes levels of security incidents and proscribes responsibilities to various utility personnel.

2.4.3 Threat Levels

Objective

Threat level advisories are assigned by the U.S. Attorney General in consultation with the Homeland Security Council, and normally apply to the entire nation. The Homeland Security Advisory System (HSAS) provides an effective means to disseminate information regarding the risk of terrorist attacks to Federal, State, and local authorities. However, water utilities should develop and implement their own threat level response plans in response to the HSAS or when it is determined that there is a more localized credible threat. A utility's plan should specify how the threat level applying to that utility will be established, e.g., automatic upon increase in the HSAS, based on an analysis of local conditions, only upon receiving a direct threat against the utility or city, etc.

The decision as to which threat condition shall be assigned involves a variety of considerations, which rely on qualitative assessment, not quantitative calculation. The higher the threat condition, the greater the risk of a terrorist attack, with risk including both probability and gravity. An important factor in the decision is the quality of the threat information itself. An evaluation of any such threat information shall include, but not limited to, the following factors:

- To what degree is the threat information credible?
- To what degree is the threat information corroborated?
- To what degree is the threat specific and/or imminent?
- How grave are the potential consequences of the threat?

Each threat condition should provide a corresponding set of proactive measures designed to facilitate decisions regarding protective actions that are considered appropriate for water utilities. The overall desire is to: reduce vulnerability; increase the public's confidence; and improve the utility's response capability during a period of increased threat. At all threat conditions, water utilities must remain vigilant, prepared, and ready to deter possible terrorist attacks.

Summary of Elements

Before developing a threat response plan, water utilities should consult with their state and local agencies to determine if these agencies have any plans providing threat level guidance. Further, water utilities in the same geographic area should consider developing joint collaboration between each other in an effort to establish a consistent response among area water utilities.

The two threat level plans presented in Appendix B present detailed actions and general protective measures for water utilities to use as guidance for the 5 threat conditions outlined by the HSAS. However, each utility should also develop its own detailed procedures to address unique activities, local conditions, and facility-specific concerns.

Discussion of Implementation Impediments

Utility employees may be affected during a response. Higher threat levels require special actions and may restrict the ability of employees to do jobs and require a higher level of security. Employees must be part of the developmental process so response procedures and protocol will be more easily accepted and followed.

Legal Considerations

Legal authorities should be consulted concerning state/local laws and regulations pertaining to the requirements of municipal agencies to terrorist threats, etc. which may affect the response procedures of the utility.

Examples of Utility Security Practices and Other Background Materials

These documents are located in section 2.4.3 of Appendix B.

A. Exhibit 1 is the Water Utility Threat Condition Emergency Response Plan For the Homeland Security Advisory System developed by the Bay Area Security Information Collaborative (BASIC).

B. Exhibit 2 is the City Of Phoenix's Threat Conditions Plan.

2.4.4 Chemical Delivery & Testing

Objective

The purpose of this section is to establish security procedures pertaining to the delivery and testing of chemicals used by water utilities. Chemical delivery could provide a potential pathway for introduction of contamination by a terrorist. Although it may not be possible to prevent all breaches of security, procedures presented in this section can help to deter adversaries from performing malevolent acts that are disguised as bulk chemical deliveries.

Summary of Elements

Generally, when ordering chemicals, the type, quantity, delivery date, and delivery timeframe should be specified. The chemical supplier should be advised that the driver of the delivery vehicle will be required to present appropriate identification (usually a Commercial Driver's License) upon arrival, and that the chemical will be tested for compliance with the order prior to allowing delivery. Also, the supplier should be required to provide advance identification of the truck delivering the chemicals, such as license plate or serial number, which will also be verified upon arrival. In essence, security practices for chemical deliveries will provide security protections through (1) knowing what type and how much of the chemical is to be delivered, (2) who is to make the delivery and when is the delivery to occur, (3) how is the chemical to be delivered and tested before it is received at the facility, and (4) training utility personnel on proper handling of chemical deliveries by following the utility's established procedures.

General guidelines applicable to good security practices include:

- All chemical deliveries and testing shall conform to established guidelines and procedures.
- If any requirement of established procedures is not met, the delivery shall not be accepted.
- All rejected loads shall be logged and reported through the chain of command.

Ordering guidelines applicable to good security practices include:

- Only authorized employees shall order chemicals for the utility.
- Chemicals shall only be ordered from vendors with a current supply contract.
- All chemical orders shall comply with contract requirements.
- All chemical orders shall specify the chemical type, quantity, delivery location, delivery driver, delivery date, and estimated time of delivery.
- The order shall be documented and require sufficient information to allow for positive identification of the vehicle and delivery driver.

Delivery and receipt guidelines applicable to good security practices include:

- Only anticipated and verified deliveries of chemicals shall be admitted to the utility's facilities.
- The delivery driver's identity and vehicle identification shall be verified and confirmed consistent with the chemical order before admittance to the utility's facilities.
- The chemicals to be delivered shall be tested per established procedures and verified consistent with the chemical order before admittance to the utility's facilities. Due to its hazardous nature, gaseous chlorine is generally not sampled and tested upon delivery. However, this material should be delivered in containers with sealed, tamper-proof locks or seals. The delivery should not be accepted if there is evidence of tampering or if the seal is broken.
- Accepted deliveries shall not be transferred until the authorized, receiving employee has verified the point of delivery, available storage and piping alignment.
- All received loads shall be logged and verification documents shall be retained with the bill of lading.

Chemical delivery information should be kept confidential. Failure to properly protect this sensitive information from disclosure beyond those who truly need it could assist an adversary with bypassing or overcoming the security procedures. The utility with excellent security procedures could be viewed as negligent in protecting against disclosure of such sensitive information, if the disclosure resulted in an injury to person or property. Therefore, open posting of information, such as driver's name, date, time, type of chemical, and so forth, should be avoided. Similarly, training of utility personnel should include warnings and procedures to be followed to guard this sensitive information from disclosure beyond those who truly need it to test and accept chemical deliveries.

Discussion of Implementation Impediments

Training and certification of personnel to prepare them adequately to perform verification of driver's identification and testing of chemicals could present a challenge depending on the number and type of personnel available to conduct the chemical testing. Another possible impediment may result from the risks associated with chemical testing and labor union restrictions on such risks.

Adopting new or better procedures for chemical delivery and testing could impact existing contractual relationships with vendors. For example, a vendor might object to a new "test first or no delivery" procedure as being outside the scope of its existing contract. This would be unusual, but possible, and might suggest use of a cooperation-building tool to avoid the potential impediment. A sample of such a tool is the second exhibit in section 2.4.4 of Appendix B – a letter to the vendor describing the security procedure and asking for the vendor's assistance and ideas.

Legal Considerations

Contracts with vendors should be well defined and make security procedures mandatory. Vendors must understand from clear contract language that failure to meet the established security procedures will result in rejection of the chemical delivery and could result in cancellation of the contract. Written contracts establishing such security procedures are preferred to oral communications or commitments from the chemical delivery vendor.

Security procedures and training established for testing and receiving chemicals should consider and address employee safety. The risk associated with handling of chemicals could lead to claims by employees in the event of injury, or to a labor dispute if an employee labor union is involved at the particular facility. Employee training and certification can reduce the legal exposure of the utility in this risk.

Examples of Utility Security Practices and Other Background Materials

These documents are located in section 2.4.4 of Appendix B.

A. Exhibit 1 is a sample operating procedure providing detailed specifications for the purchases of water treatment chemicals. Specifications are provided for bid package requirements through delivery of the chemicals to the water treatment plant.

B. Exhibit 2 is a sample of chemical delivery procedures incorporated into a letter from a water utility to its chemical suppliers. This letter shows one way to implement such procedures in a cooperative way by focusing on safety concerns of both the supplier and the water utility.

C. Exhibit 3 is a sample of a water utility's procedure to establish security standards for ordering, delivery, and receipt of bulk chemicals and fuel at its facilities. This sample shows one way to establish scope, responsibilities, and guidelines for this security topic.

D. Exhibit 4 is a sample of security guidelines used by a water utility to provide uniform security inspection requirements for all bulk chemical deliveries to its facilities. This exhibit also includes a sample Chemical Delivery Inspection Form that indicates the types of information that may prove useful in tracking bulk chemical deliveries.

2.4.5 Package & Mail Delivery

Objective

The purpose of this section is to minimize handling and prevent internal distribution of suspicious mail, parcels, and small packages, which could pose a health or security threat to water utility personnel. The episodes of anthrax contamination of mail in 2001 confirm the real threat of this technique of terrorism.

Summary of Elements

An effective mail handling procedure should be clearly defined and periodically reviewed with water utility personnel. Specifically, the following four components are examples of an effective mail handling procedures. First, restrict the times and locations of package and mail deliveries, and post the schedule for deliveries at each entrance or reception area. Second, prevent the mixing of external and internal packages until trained personnel have thoroughly inspected and cleared each item for final delivery. Third, designate and train personnel to perform initial screening inspections of mail prior to final delivery. Fourth, establish mail-screening supervisors to coordinate local mail, parcel, and package inspections; to provide confirming or second opinions to inspectors who perform the initial mail screening; and to decide when to activate emergency response procedures.

A designated area within each mailroom should be segregated for initial sorting and inspection of incoming mail to prevent potentially harmful items from leaving designated receiving and inspection areas. Internal mail may also need to be inspected in a similar fashion. In any case, internal mail and incoming mail should not be mixed until after the screening process is complete. The U.S. Postal service can provide extensive information on safe mail handling procedures.

Discussion of Implementation Impediments

Increased security measures may require additional personnel or may require existing personnel to share new responsibilities associated with increased security for package and mail delivery. Furthermore, additional facilities may be required to adequately separate mail sorting areas from other facilities. In any case, the increased security measures may require expenditures to hire additional personnel or to construct additional facilities.

Legal Considerations

Personnel who are responsible for screening and sorting mail could be subjected to increased risk. The result could be that the employer is subjected to increase liability.

Examples of Utility Security Practices and Other Background Materials

These documents are located in section 2.4.5 of Appendix B.

A. Exhibit 1 is a mail and package handling procedure from Massachusetts water Resources Authority.

B. Exhibit 2 presents the mail handling procedures of the City of Phoenix Water Services Department.

2.4.6 Guard Service and Use of Deadly Force

Objective

The use of guard service is routinely designed to provide the utility with protection during and after work hours, weekends and holidays. To supplement routine and normal utility security, a guard service should provide an extra set of eyes that control access at the gate and patrol the perimeter for any attempted intrusions, and/or to supplement security at the utility. Using a guard service enables utility employees to address daily operations while guard service personnel address utility security. Costs are an issue for consideration in determining whether to use a dedicated guard service or local law enforcement personnel. The objective of any type of guard service should be a deterrent to and prevention of intrusions at the utility, but a guard service can meet these general objectives in a number of ways. The use of deadly force should be carefully considered regarding when use of deadly force may be appropriate, and who may use it. First and foremost, standard operating procedures and utility policies should be developed and to define the purpose and responsibilities of any guard service. These policy and procedures can have significant impacts in improving overall water facility security.

Summary of Elements

Several issues must first be addressed to develop and establish utility policies and procedures in support of guard service. The utility should review the following issues and determine what the most appropriate course of action for the specific utility is:

- Guarded gate versus facility perimeter patrol
- On-foot versus vehicular patrols
- Armed versus unarmed personnel
- Guard service versus local law enforcement for increased surveillance and patrolling of utility property.

The utility must first identify the area to be patrolled, whether it is the front gate and/or the facility perimeter, on-foot or vehicular patrols. The size of the utility may determine the area to be patrolled. The volume of traffic to the facility (workers, contractors, vendors, and visitors) will also assist in determining use of guard service. Guard service is variable throughout the country, and often employs a less than adequate work force. Throughout industry, use of guard service is most often relegated to unarmed personnel (i.e., no weapons other than a baton and pepper spray) who provide surveillance as a deterrent to and prevention of intrusion, working closely with local law enforcement should use of deadly force be necessary. A weapon at a facility in untrained hands can be a costly mistake. Accidents do happen and liabilities are large in these types of incidents. When the utility has decided upon the above issues, policies and standard operating procedures should be developed, established and communicated.

Whatever type of guard service the utility decides upon, all workers should be trained in their roles and responsibilities, and knowledgeable of the utility policies and standard operating procedures. A hired guard service agency should meet the following personnel requirements:

- Trained personnel who have undergone employee background checks by the employer
- Personnel who are familiar with chain-of-custody and securing an investigation site procedures
- Knowledgeable of alerts on potential terrorist activity correlated with the Department of Homeland Security's color-coded advisory system
- Trained on habits/patterns of behavior that might indicate surveillance of facility taking place
- Provision of all tools, supplies, communication systems, vehicles, etc. necessary to conduct utility surveillance
- Provided with information on emergency preparedness and response resources
- Clearly identified (uniformed and badged) personnel

Patrol parameters:

- Scheduled, but at varied times security surveillance patrols
- Defined area of patrol

This method of surveillance allows for a varied routine that does not occur at the same time daily. The reason is that anyone observing the utility cannot predict when the patrol will occur. Terrorists do not like changes in routine or non-patterns; they will typically move to other more predictable targets.

Communications and record keeping policies and standard operating procedures are critical to the success of a security guard program. The utility should work with the guard service to establish security practices including the following:

- Communications and record keeping:
- Utilization of record keeping activities to record and report on physical vulnerabilities and security solutions;
- A secure means for reporting security incidents;

- Communication systems established and maintained with local law enforcement for arrests and force
- Use of communication tools (i.e., radios, cell phones, land lines, PDAs) with scheduled reporting times and monitoring by a central command.

For example, if the guard on his/her security surveillance patrol notes a fence break or open gate, this information would ideally be reported to the utility via established record keeping thereby identifying a security issue easily remedied. As well, should an intruder be detected, established and immediate communication to central command or local law enforcement would provide immediate response to the utility. The need for this timely communication could prevent loss of life, injury, and utility sabotage.

It is incumbent on the utility to provide the guard service direction and information critical to meeting utility objectives. This should include:

- Utility information tools and direction;
- Updated emergency contact lists;
- Participation of security personnel, including guard service, in utility emergency preparedness and response exercises;
- If the guard is stationed at the gate, access to regularly updated lists of authorized personnel (i.e., workers, delivery, vendor and contract personnel) with standard operating procedures for access and egress
- Established enforcement procedures for strict access, parking and delivery procedures for contractors, vendors and visitors.

Discussion of Implementation Impediments

The issue of force, and specifically deadly force, is a very sensitive topic. Concerns include accidents, untrained individuals, definition of activities that prompt deadly force, liabilities, local law enforcement rules and regulations, and weapons on site. For these reasons it is important for a utility to determine whether to use a guard service or to rely upon local law enforcement to address deadly force. It is advisable to discuss this issue with local law enforcement before a final decision is made by the utility.

Again, costs are an issue for consideration in support of guard service. In light of initial and improved access controls such as fences, locks, perimeter lights, and intruder alarms at critical utility assets, guard services must also be included in the equation. Each utility must determine whether to use a dedicated guard service or local law enforcement personnel.

Legal Considerations

Federal, State and local laws and regulations apply to detaining and imprisoning individuals, injury and use of force and/or deadly force. These considerations as well as liability issues must be assessed by utilities in their decision concerning use of guard services, parameters for behavior by the guard service, and how to integrate security surveillance programs with local law enforcement.

Examples of Utility Security Practices and Other Background Materials

These documents are located in section 2.4.6 of Appendix B.

A. Exhibit 1 is the Security Guard Protocol developed by the City of Phoenix Water Service Department. This document provides a framework for use of a security guard service.

2.4.7 Construction Site Security

Objective

As utilities are often expanding, upgrading, or performing large construction and maintenance projects, much of their property may be considered a construction site. Activities within and surrounding the construction site may make it necessary to modify or temporarily suspend normal security procedures. This section provides guidelines and examples as to how utilities may mange security at their construction sites.

Summary of Elements

As much as possible, security elements surrounding the construction site should be *at least* equal to those areas of the utility's facility when construction is not taking place. For example, if lighting is normally available at the facility, then lighting should be available and provided for the construction site. Essentially, any construction operation becomes part of the facility and should be treated as such. Without equivalent security, a potential intruder can possibly gain access to the facility through the less protected construction area.

In cases where the construction operation is large, there may be more employees and deliveries to the construction site than during normal facility operations. It is possible that this increased activity may require a higher level of security at the facility than normal. The following issues have been identified as critical in such a policy.

- Security of the utility must be included in the contract with the contractor operating the construction site.
- Access to the construction site must be controlled by badging.
- Consider whether construction workers should be subjected to background checks, particularly if the construction workers will have access to sensitive facilities.
- Contractor must notify the utility of deliveries to construction site, particularly if the delivery will enter through the main facility gate.
- A schedule of construction site activities should be provided to the utility.
- Parking for construction site employees and visitors must be supplied and monitored, preferably outside the facility perimeter.
- The construction site should be fenced.
- Where possible, create a separate entrance for construction site only.
- A policy should be established for pre-approval of construction site activities outside of normal utility working hours.
- Where it is impractical to provide the same level of physical security around a construction site as around the facility itself, adequate security should be installed between the facility and construction area.

Discussion of Implementation Impediments

Additional security or security constraints required for construction sites will increase the construction contractor's costs, and this increase is likely to be passed along to the utility. For example, if a utility requires that its contractors perform background checks on its employees, the contractor may incur an additional cost depending on whether the contractor has already performed such checks.

The amount of activity at a construction site may vary daily, leading to increased traffic of both people and vehicles. It may not be practical or possible to have every contractor and sub-contractor, and their respective employees, properly badged before entering the construction site. Therefore, consistent and ongoing efforts to require and monitor proper badging of the construction workers must become the ultimate goal.

Space and access constraints may make it impossible to have a separate construction site entrance. Moreover, construction equipment, materials and supplies within the construction site may present additional security risks not necessarily considered in a particular utility's normal security plan. For example, the construction site may include storage of large quantities of chemicals used for concrete treatment, large quantities of paint, or even bags of cement – all of these could be used to cause damage to the utility or its operations. Another concern might be that vibrations from large equipment could be severe enough to set off security alarms, or damage sensitive utility operations monitoring equipment.

Legal Considerations

Contracts should specifically state the utility's requirements for construction site security. As with any contract, language should be complete, clear and concise to avoid claims and hidden costs. Care should be taken to ensure that the rights of construction site employees are not infringed upon by the security policies of the utility. Similarly, care must be taken to protect the employees and customers of the utility from unauthorized or undesirable individuals working at a construction site of the utility.

Examples of Utility Security Practices and Other Background Materials

These documents are located in section 2.4.7 of Appendix B.

A. Exhibit 1 is a good representation of a construction site security policy from the City of Phoenix.

B. Exhibit 2 is a specification from a water utility plant upgrade at Alameda County Water District.

2.4.8 Neighborhood Watch On Remote Facilities

Objective

Neighborhood Watch is a national crime prevention program designed to reduce neighborhood crime through observant citizens. The concept can also be applied to water utilities. All utilities have unstaffed, remote facilities. Organized neighbors can assist utilities in protecting these facilities from vandals, criminals, and terrorists. This involves citizens taking a responsibility for safeguarding their water supply by becoming alert and attentive to changes in their neighborhood. Citizens can report crime in progress or suspicious activity to the water utility giving accurate information that could assist in preventing a malevolent act or apprehending a potential intruder. By devising effective neighborhood watch policies and implementation, water utilities can enhance their property protection and strengthen customer relations.

Summary of Elements

A number of utilities have established neighborhood watch programs. The process used include the following steps:

- Mailings or other communications sent to residents requesting their participation in the neighborhood watch program. If possible, the utility should hand address and stamp the mailing to personalize the interaction with community members. A letter detailing the program and their valued assistance should accompany a postage-paid response card. The response card should provide an opportunity for residents to agree or decline to participate in the program and mail back to the utility (see Exhibit 2).
- The utility may want to concentrate on residents who live near unstaffed major facilities. However, since every fire hydrant is a potential pathway for contamination, a service area-wide program would also be worthwhile.
- Upon receiving confirmation by residents of their participation in the program, a brochure should be mailed to these residents detailing how the volunteers can assist water utilities protect their water supply, provide examples of suspicious activities and familiarize them of normal utility procedures and actions.
- Examples of suspicious activities include:
- Unknown vehicles and persons around the facility,
- Persons seen videotaping, photographing, or sketching drawings of a facility,
- Persons seen spray painting or tampering with any equipment at a water facility,
- Non-utility personnel inquiring about a facility (in-person or by phone),
- Open gates or doors at the facility with utility personnel not present, and
- Any activity or noise observed at the utility during non-business hours.
- Examples of what to provide volunteers concerning normal utility procedures and actions include:
- How to identify an official utility vehicle (vehicle markings, color, etc.),
- Utility employee uniform descriptions (color, logos, etc.),
- Open/close gate procedures,
- Normal time spent on site visits,
- Utility employee identification, and
- Utility phone contact information.
- Procedures for reporting suspicious activities and malevolent acts should be provided to participants, including contact information, information to report, and information the caller will be asked to provide to the utility, etc. Utility dispatchers should ask for the street address to identify the facility, a description of what was seen or heard, and

a callback number. The caller should be assured that their identity will not be publicized.

- Utility business cards and/or a refrigerator magnet containing utility contact information can also be included in the mailing packet to participating residents.
- Scheduled watch meetings with program participants can strengthen the relationship with the community as well as provide an opportunity to gain valuable insight and ways to improve the program.
- Coordination with municipal and county law-enforcement agencies should also be pursued concerning their neighborhood watch programs.

Discussion of Implementation Impediments

Citizens must be instructed not to take action themselves if they see what they consider a criminal act in progress. Rather, the citizens must call the utility or 911. The program should not be seen as encouraging vigilantism.

Legal Considerations

Legal staff should examine the liability of the utility should a citizen be injured or killed while trying to stop an incident, even if they have been advised not to do so.

Examples of Utility Security Practices and Other Background Materials

These documents are located in section 2.4.8 of Appendix B.

A. Exhibit 1 is a package of materials used by the Alameda County Water District for their neighborhood watch program, "Water Watch", including a reply card and instructions.

B. Exhibit 2 is a letter San Antonio's BexarMet water utility sends to its customers to gather support for their neighborhood watch program, "Operation Eagle CLAW: Citizens Looking After their Waterworks."

2.5 COORDINATION (INTERNAL & EXTERNAL)

2.5.1 Coordination: Public and Media Relations

Objective

The purpose of this section is to provide policies and procedures for the coordination of communication between the organization, the media, and the public in the event of an emergency situation. The objectives include:

- To standardize a process of determining when public notification is required.
- To establish a protocol for media and public announcement.
- To standardize how public and media announcements are made.

Summary of Elements

Effective communication with the media and public during a crisis can restore or maintain public confidence, avoid panic, and disseminate critical emergency information. Ineffective communication with the media and public may turn a relatively minor incident into unnecessary loss of public confidence, public panic, and generation of conflicting rumors as to what has happened and what people should be doing to ensure safe drinking water. An effective public and media crisis communications plan includes the following elements:

- Identify a media and Public Information Officer (PIO) and Emergency Communications Team (ECT).
- Establish written rules and protocols for the PIO and ECT.
- Pre-determine how information will be released in case of an emergency.
- Identify which media organizations will be notified.
- Develop a written plan for conducting news conferences, including location and personnel in charge of news conference.
- Prepare a written plan for disseminating information to employees.
- Develop a written policy of media relations during emergency situation and during non-emergency situations.
- Before an incident, develop fact sheets about the utilities operations and roles in the community.
- Before an incident, develop fact sheets on how the public should treat water during a given emergency, including a list of top contaminants and information about the contaminant. To ensure a clear message this information should not be released until required by an emergency, and then only the relevant fact sheets should be disseminated to the public.
- Before an incident, develop press release templates.
- Maintain phone lines dedicated to customer service and prepare scripts for operators.
- Train personnel on how to treat media.
- Provide personnel with cards containing contact numbers and names of designated spokesperson.

- Conduct specialized training provided for designated spokesperson(s).
- Conduct exercises with a media relations component.
- Develop a written procedure for post event analysis, review of media reports, making clarification and corrections to media reports.

Discussion of Implementation Impediments

Media training should be required for all employees, as emergency situations may force any employee into a media confrontation, but this training is an additional cost to the utility. Utilities may feel that money spent on media training could be better spent on resources and equipment to prevent or mitigate emergencies. Employees may be unwilling to accept the role of interacting with the media.

Legal Considerations

The accuracy of information given to media services could be an item for legal consideration, including liability for the incident. Typically, the following information should not be given to the media:

- Speculation of any type, particularly on the cause of the emergency.
- Cost estimate of damage.
- Information about insurance coverage.
- Premature assessments of the performance of response personnel.
- Implications of negligence or blame.

Examples of Utility Security Practices and Other Background Materials

These documents are located in section 2.5.1 of Appendix B.

A. Exhibit 1 from the AWWA is an excellent, but brief overview of a media relations policy and a checklist. A prospective user could use this as an outline and input specific utility information.

B. Exhibit 2 is a condensed version of a crisis communications plan from a major Western utility.

C. Exhibit 3 is a press release from the Alameda County Water District, *Alameda County Water District Responds to Increased Threat Level.*

2.5.2 Coordination: Mutual Aid and Regional Cooperation

Objective

The purpose of this section is to provide policies and procedures for regional emergency preparedness, disaster response, and mutual assistance matters for public and private water and wastewater utilities. Participating agencies in the mutual agreement would assist one another with emergency planning, response, and recovery information before, during and after an emergency.

Summary of Elements

Effective mutual aid and regional cooperation agreements should encompass the following:

- A mutual assistance agreement and process for sharing emergency resources among members.
- A mutual assistance program consistent with other statewide/regional mutual aid programs.
- The resources to respond and recover more quickly from a disaster.
- A forum for developing and maintaining emergency contacts and relationships.
- A process to review lessons learned from disasters.

The core of a mutual aid agreement should be:

- An emergency equipment database that matches utility resources to a member's needs during an emergency. A member can locate emergency equipment and trained personnel that they may need in an emergency.
- The willingness of member utilities to respond to other member's emergencies.
- Reimbursement provisions.

The Bay Area Security Information Collaborative (BASIC) in California is an excellent example of regional cooperation in security planning. An example of BASIC's threat level response protocol is presented in section 2.4.3 on Threat Levels. Also in California, the Water/Wastewater Agency Response Network (WARN) provides a model for internet-based mutual aid.

Discussion of Implementation Impediments

Execution of mutual aid may pose problems if member utilities are not fully committed to the agreement. Use of Internet resources, i.e. password protected databases, may provide the best forum for information and idea exchange. Cost of database maintenance may be assumed by the larger utilities, so that the smaller utilities will not be financially encumbered by agreements that benefit all utilities. Each utility would be responsible for maintenance of its own information within the database.

Legal Considerations

Mutual-aid protocols should be documented by a legal agreement that sets forth the responsibilities of all signatory agencies. This agreement should specify the signatory agencies; definition of situations in which it will be implemented; an organization for managing the agreement; procedures for provision of assistance; reimbursement policies; and liability, hold harmless, indemnification, and workers' compensation provisions.

Examples of Utility Security Practices and Other Background Materials

These documents are located in section 2.5.2 of Appendix B.

A. Exhibit 1 is a description of the BASIC group.

B. Exhibit 2 is the agreement forming the BASIC group.

C. Exhibit 3 is a comprehensive description of the CALWARN mutual aid process.

2.5.3 Incident Reporting, External Notification, and Response

Objective

The purpose of this section is to provide policies and procedures for the reporting of incidents and coordination of communication within the organization and with emergency response agencies. The objectives include:

- To factually assess the situation and determine whether plan activation (level of response) is warranted.
- To prescribe action to:
  - Notify police, fire, paramedics or other emergency response entities.
  - Communicate facts about the emergency internally and externally.
  - Notify internal employees or external contractors with the knowledge and skills to implement a solution to the problem or mitigate the effects of the incident.
  - Restore normal operations and confidence.

Summary of Elements

A water utility's security practices should establish a procedure for all water utility employees to follow when any suspicious incident occurs. It should identify points at which an employee should notify superiors or emergency personnel (e.g. 911) immediately. The written procedure should identify specifically how to report an incident, specifically who to notify and also what the response procedure should be. For instance, in the simple example of an employee finding a building door broken into. A written policy should address what the employee should immediately observe (time, people in area, other suspicious evidence, packages), who the employee should call (local police and then supervisor, or visa versa), how the employee should respond (keep watch on door, do not enter, until local police arrive). The bullet list below was developed for some of the more serious aspects of such a policy:

- Emergency telephone numbers listed and readily available to all employees, posted on walls near phones, etc.
- Written "what to do" policies for the following:
  - Power outage
  - Natural disaster (earthquake, tornado)
  - Verbal threats (bomb threats/physical threats)

- Alarms (SCADA or door/ window – non SCADA)
- Vandalism
- Physical Violence (Assault)
- Suspicious persons
- Chemical or Biological contamination
- Chemical Release (Chlorine)
- Burglary/Theft
- Fire
- Medical Emergencies
- Report forms to report threats/ incidents.
- Follow up or lessons learned.
- Procedures or schedules for performing drills.
- Flowcharts are recommended.

Discussion of Implementation Impediments

Continual training may be necessary to train new employees and procedures may require updating as new threats become apparent. Many utilities may struggle with the costs of implementing and maintaining a system. Considerable effort may be required to contact and create relationships with emergency response personnel. Live drills may be the most effective way to train employees and emergency response personnel but this may be cost prohibitive.

Legal Considerations

None noted.

Examples of Utility Security Practices and Other Background Materials

These documents are located in section 2.5.3 of Appendix B.

A. Exhibit 1 provides a good example of "what to do" procedures. Also included are specific emergency numbers and report forms for incidents such as bomb threats.

B. Exhibit 2 includes excellent and very specific flowcharts for an intruder alarm incident and a water contamination incident.

C. Exhibit 3 is a response checklist.

D. Exhibit 4 is a typical bomb threat call checklist from Phoenix, which could be used for other threats as well.

E. Exhibit 5 is a bomb threat response procedure from Phoenix, which could be used for other threats as well or as a model procedure for incident reporting and emergency response notification.

F. Exhibit 6 is a checklist to record threats against a utility. It issues guidelines of how to deal with threats made through various methods such as phone, e-mail or in person.

2.5.4 Coordination: Health/Medical and Pharmacies

Objective

The purpose of this section is to provide guidance for coordination between the health/medical community and water providers. Primary care practitioners throughout the United States must be especially vigilant in light of the fact that they are likely to be the first to observe unusual illness patterns and must therefore understand their critical role in protecting water resources and their community's health. Any future strategic plan to maintain and protect water quality and safety in the United States, particularly for intentional contamination, must include physicians as stakeholders and active participants in this ongoing public health challenge.

Inaccurate assessment of the prevalence and incidence of waterborne disease and slow recognition of a waterborne outbreak may limit the successful implementation of water quality treatment and protection strategies by local water utility practitioners and local health authorities. Misdiagnosis and under diagnosis of waterborne disease by the medical community may result in significant morbidity and mortality, particularly in vulnerable populations at increased risk of disease as a result of exposure to waterborne pathogens and chemical contaminants. The increased risk of morbidity and mortality results in a need for practicing healthcare providers to recognize unusual disease trends and early warning signs that may result from potential biological or chemical terrorism.

Local pharmacies may notice an increase in sales of over the counter and prescription drugs to treat gastrointestinal illnesses, which could be an indicator of water contamination.

Summary of Elements

An effective security practice should include the following two elements:

- Reporting any unusual disease trends that may possibly be related to waterborne exposure (even prior to laboratory confirmation) will allow public health interventions to be planned and implemented more effectively. In addition, local and state public health agencies offer significant resources of updated information regarding many aspects of waterborne pathogen exposure and disease characteristics for both practicing clinicians and their patients. Many state public health agencies post sophisticated Websites with detailed information regarding specific precautions and risk reduction activities for waterborne pathogen exposure for use by individuals in their community.
- Communication with local water utility practitioners also improves water protection efforts such as ongoing monitoring of infectious pathogens in water reserves, as well as notification of any possible penetration of water treatment facilities. Water utility practitioners also act as resources of information to medical practitioners for information regarding water quality monitoring and treatment protocols for both the healthcare provider and their concerned patients. In addition, physicians must work

in collaboration with those responsible for protecting the safety of drinking water reserves from possible threats of bioterrorism and infrastructure assault.

- Coordination with local pharmacies could provide a water utility with an early indication of potential contamination through an increase in sales of drugs for gastrointestinal illnesses. Some pharmacies may have new point-of-sale inventory control that could give nearly instantaneous indication of an increase in sales.

The New York City Department of Environmental Protection has developed a very sophisticated system to monitor gastrointestinal illness through communication with the medial community and pharmacies. For more detail on this program, the reader is directed to the DEP website: www.nyc.gov/html/dep/html/wdrap.html.

Discussion of Implementation Impediments

Most utilities are currently not working with the medical community unless a specific issue arises. Medical health departments are relevant partners in light of recent events. Utilities need to consider the medical community in all aspects of protecting water resources and their community's health. Information sharing and coordination with the medical community are areas to be improved upon.

Recognizing and treating waterborne disease and the health effects of acute and chronic exposure to water pollution is a diagnostic dilemma for the majority of practicing physicians in the United States. Challenges associated with accurately diagnosing and managing waterborne diseases and chronic exposure to waterborne contaminants include but are not limited to the following scenarios:

- Many of the signs and symptoms of waterborne disease and the health effects of water pollution are non-specific and often mimic more common medical conditions and disorders.
- Patients may not be aware of previous waterborne exposure and may not provide an accurate exposure history. In addition, many of the pathogens and chemical contaminants found in the water environment are not unique to water and may exhibit multiple routes of exposure and result from other sources of environmental contamination.
- Waterborne exposure events in a healthy patient population often present as benign symptoms or self-limited illness while the same waterborne exposure events in a vulnerable patient population may result in significant morbidity including chronic and life-threatening disease.

Partnering with the medical community can raise awareness of waterborne threats and increases the likelihood that waterborne exposure can be identified early. For example, AWWA collaborated with Arnot Ogden Medical Center in Elmira, N.Y., to develop an on-line resource to provide medical professionals with critical information that will assist physicians in better diagnosing symptoms of waterborne diseases in patients. By working together, the water and medical communities were able to develop a resource that facilitates early diagnosis and appropriate treatment of waterborne disease.

Legal Considerations

Failure to establish effective response coordination with the health/medical communities could create liability on a water utilities part in the event of a contamination incident.

Examples of Utility Security Practices and Other Background Materials

None

The Centers for Disease Control and Prevention (CDC) and the Environmental Protection Agency maintain a collaborative system for collecting and reporting waterborne disease outbreaks from both microbial and chemical origins. This surveillance system collects data on outbreaks and contamination events for both drinking and recreational water in the United States. Detection and investigation of these events is primarily the responsibility of the local, state and territorial public health departments with reporting to the CDC.

Enhanced communication with local water utility practitioners will also be critical in the future in order to improve water protection efforts such as ongoing monitoring of contaminants in water reserves as well as notification of any possible penetration of water treatment facilities.

2.6 TECHNICAL AND CONTAMINATION RESPONSE

2.6.1 Employee Safety

Objective

The main objective of this security procedure is to provide procedures and recommended responses designed to provide employees with adequate protection, to ensure employee safety, or at least to minimize the health threat to employees during actual contamination events.

Summary of Elements

Following is a list of important elements:

- Personal Protective Equipment (PPE),
- Employee Safety Training,
- Standard Operation Procedures (SOP),
- Field test kits for water sampling,
- Emergency response protocol,
- Monitoring response protocol,
- Linkage to Health Departments,
- Established Material Safety Data Sheets (MSDS) access,
- EPA peer review of procedures.

As shown in Appendix B – 2.6.1 Exhibit 1, PPE is divided into four categories based on the level of protection provided by the protective equipment (Levels A-D). The four protection levels provide guidelines for the selection of appropriate PPE. Situations may warrant the use of a combination of PPE selected from the different protection levels to provide appropriate employee protection from the hazards associated with the work performed.

- Level A—the greatest level of skin, respiratory, and eye protection is required.
- Level B—the highest level of respiratory protection is necessary but a lesser level of skin protection is needed.
- Level C—the concentrations and types of airborne substances are known and the criteria for using air-purifying respirators are met.
- Level D—a work uniform that affords minimal protection and used for nuisance contamination only.

In considering how to develop an employee safety protocol, it is helpful to consider the procedures of other public utilities. A good example protocol is found in the procedures of a large water utility in the western United States. This anonymous utility allows personnel to respond to Level C situations, provided that a respirator is adequate PPE for respiratory system protection. If a respirator is not adequate, then a HAZMAT crew should be used in either Level A or Level B PPE, as appropriate.

Employee safety training is the cornerstone to Employee Safety. All employees charged with roles and responsibilities related to contaminant monitoring and/or response to contamination, should be trained in not only PPE, but also HAZWOPER and related hazardous material handling courses. It is critical that initial training be accompanied by "refresher" courses and appropriate certification (i.e., HAZWOPER 40-hour, Supervisory, etc.). These courses are offered regularly through Federal (FEMA Emmetsburg Training Facility) and private sector offerings. Training records should be maintained and updated for employees. In certain cases, contractors are also required to conduct sampling operations. In this case, contractors should be required to be trained in all appropriate HAZMAT and PPE procedures, and this training should be documented. The contracting firm should be required to maintain all training records, and conduct refresher courses as needed on an annual basis.

In terms of standard operation procedures (SOP), and field test kits for water sampling, industry standards and checklists should be utilized as much as possible to eliminate accidents and exposures. Checklists should be developed by the individual utility in support of how the sampling should be conducted, employee safety procedures employed, PPE required, notification requirements, and response recommendations. OSHA and EPA's "Superfund" sampling protocols can be used as a guide to develop such procedures and checklists.

Emergency response and monitoring response protocols are a critical element to employee safety. How a utility responds to a known event, and specific monitoring response protocols used to determine whether a contaminant was introduced, will determine a utility's overall safety program. Again, utilization of OSHA and EPA's emergency response rules and regulations can contribute to the development of an excellent response program. Utility established monitoring response protocols could follow EPA's Chain-of-Custody procedures as a resource.

Linkage to local Health Departments, and established access to Material Safety Data Sheets (MSDS) is integral to employee safety. Should an employee be exposed to a particular toxin, chemical or biological agent, an established relationship with a local Health Department could prove beneficial. A contact list (included in the utility Emergency Response Plan), including the appropriate (and back-up or alternate) Department of Health contact, ensures rapid response to employee safety issues. As well, access to MSDSs will ensure accurate exposure and response information on the contaminant is available.

EPA and OSHA have perhaps been the leaders in terms of information in support of employee safety. It would be useful for any utility, regardless of size, to utilize EPA and OSHA's peer review of procedures and training programs in support of employee safety.

Discussion of Implementation Impediments

None found.

Legal Considerations

Contamination response procedures should be coordinated with Federal, State, and local agencies to verify the quality of the proposed technical and contamination response procedures against Federal, State, and local legal requirements. Chain-of-Custody procedures for sampling and monitoring requirements should be included in all standard operating procedures. OSHA and EPA HAZWOPER training requirements should be included in all hazardous material sampling and response programs and documented in terms of assigned roles and responsibilities.

Examples of Utility Security Practices and Other Background Materials

These documents are located in section 2.6.1 of Appendix B.

A. Exhibit 1 is entitled General Description and Discussion of the Levels of Protection and Protective Gear, Safety and Health Regulations for Construction, Occupational Health and Environmental Controls. This OSHA document describes the different protection levels (Level A-D) and gives general guidance for when each level of protection should be implemented.

B. Exhibit 2 is entitled Compliance Guidelines, Safety and Health Regulations for Construction, Occupational Health and Environmental Controls. This OSHA document illustrates generally safety compliance guidelines regarding contact with hazardous materials. Specifically, these guidelines could be used to develop an employee safety program for an individual utility.

2.6.2 Contamination Response Including Threat Validation Procedures

Objective

The purpose of this section is to outline the steps and procedures to be taken by a water system that has experienced a threatened, suspected, or actual intentional contamination/intrusion into their public water supply system/facilities. This is a very complex and sensitive issue and many utilities are developing response protocols. The main objectives of a contamination response plan include validating that a threat is credible or that a contamination event has occurred and defining recommended responses to both threatened and actual contamination events

Summary of Elements

Reports of water supply contamination have the potential to create public panic, even if no contamination has actually occurred. This fact makes a false threat of contamination an effective and simple way to disrupt a water utility. The utility must be able to quickly verify if contamination has or has not occurred. An effective verification process will identify means and methods for determining if a contamination has occurred. This includes effectively assessing the threat and laboratory response plans.

The process may also identify "triggers" which will define when emergency procedures for dealing with a contamination event go into effect, even if a contaminant has not been detected. These "triggers" may be based on time lapse, preliminary analysis, local or federal law enforcement information, where the contaminant threat is located, or observations of utility employees. For example, if local or federal law enforcement gives the utility information that an attack is imminent, that could trigger the utility to bypass validation procedures and employ emergency measures immediately upon receiving a threat, alarm or other evidence. Other triggers for immediate action might be a reservoir intrusion alarm or finding empty containers near access points to the water.

CDC, EPA, and other Federal agencies are identifying specific contamination agents of concern. A variety of illness causing chemical and biological agents are clearly of concern to utilities. In addition, materials, such as petroleum products and insecticides, while not necessarily a major threat to human health, can cause enormous disruption and cost to the utility for clean up. Individual utilities should use this information, once available, rather than trying to develop their own list of contaminants and procedures for responding to those individual contaminants.

Contamination Event Detection

A contamination event can be detected and verified through various pathways, which will determine specific response protocols. Contaminant event detection pathways include, but are not limited to:

Intrusion

- Alarm or sensor with confirming capability (e.g., at least two levels of sensors) indicates someone (or something) at the facility.
- Utility employee or law enforcement personnel observe unauthorized person at a facility.
- Utility employee or law enforcement personnel observe suspicious evidence at a facility (e.g., open barrel of chemical).

Public or Customer Reporting

- Customer phones in an unusual taste/odor/color report.
- A suspicious event is reported at a water facility (including hydrants).
- A potential contaminant is observed (visual or odor) and reported in the source water.
- A governmental agency or other water utility reports a contaminant in the source water.

Routine Water Quality Monitoring

In this pathway (e.g., grab sample and on-line), the background conditions have been determined adequately to allow assessment of any change that can be interpreted as potential contamination vs. changes in flow or sources, such as:

- On-line chlorine monitor reports an unusual drop in residual in the distribution system.
- On-line pressure gage indicates a sudden drop in pressure in distribution system.
- Other on-line water quality monitors indicate possible contamination in the distribution system.
- Routine grab samples in source water or distribution system suggest a possible contamination event.
- Source water monitor indicates the presence of an unusual contaminant.
- Biological tools in use by the utility, i.e., Daphnia water fleas, phosphorescent bacteria, fish tanks, etc., indicate unusual water quality. These can provide an

indication that a foreign substance is present, but will not typically identify the toxic agent.

Utilities should determine an acceptable water quality standard based on operating history. Any deviation from that standard should then trigger an immediate investigation.

Medical Reporting

- An emergency room reports unusual amount of activity that may be associated with a waterborne disease.
- Health Department reports unusual amount of activity that may be associated with a waterborne disease.
- Pharmacies report unusually large sales of a drug that may be associated with a waterborne disease.
- Schools report unusually large absence rate that is not explainable by normal flu or virus.

Threats

- FBI/local law enforcement reports a credible threat to the water system.
- A threat is communicated to the water utility (by phone, letter, email).
- A possible threat by an insider is received or suggested.
- Media suggest that here may be a threat to the water utility.

Contamination Response

Following the identification of a suspected contamination event, there are several simultaneous or near simultaneous actions that a utility should take:

- Isolation of the portion of the system suspected to be contaminated.
- Initial notifications and responses.
- Assessment of the event through on-site inspection and sampling, and a decision on whether this is a credible/serious contamination event.

Once a contamination event is confirmed, various actions are taken including wider notification; mobilization; and isolation, containment and mitigation of the contamination. Response procedures need to be both utility-wide and site/facility specific.

As soon as a contamination incident is suspected, utility equipment should be operated to isolate any potential contaminant, thus containing the event's undesired consequences to the specific location of contamination. Isolation minimizes the likelihood of widespread impact from the contamination. For this reason it is important for utilities to be able to isolate sections of the system. Once the suspected contaminated area is isolated, sampling can take place to determine if, indeed, contamination has occurred.

As the response gets underway, a variety of agencies and individuals will become involved. These can include State and local health and public utility agencies, the EPA regional office, police and the FBI, local and State emergency management agencies, hazardous materials response units, and elected officials. If the situation is serious enough, the Mayor or Governor may be asked to issue a declaration of a State of Emergency, which could trigger the activation of additional resources, such as the National Guard and military Civil Support Teams. A Federal Emergency or Disaster Declaration is also possible in a catastrophic situation, making federal resources available. The multiplicity of organizations involved and the resulting potential for chaos and conflicting control demands that utilities thoroughly coordinate response plans. The Local Emergency Planning Committee and other area utilities are probably good points to start the coordination process. Note that the orderliness of the response will be enhanced if the utility adopts the Incident Command System (ICS) as their internal response structure, since most other agencies operate with an ICS organization.

The specific contaminant may not initially be of concern in determining the immediate decisions the utility will take. Rather, a procedure to determine the contaminant follows on-site inspection, where additional evidence might be found, such as empty containers. As a general precaution, the contamination is presumed to be serious and have potential for a major public health impact. Ultimately, a variety of resources are available to assist a utility in determining if a contaminant is present and, if so, what it is. Most state health agencies are developing sampling and laboratory testing capabilities for agents of concern. EPA, CDC, and the FBI also have sampling and testing capabilities. Local hazardous materials units may be able to provide assistance. It is recommended that utilities consult with local and State health agencies and local hazardous materials agencies to identify available sampling and testing resources, capabilities, procedures, and requirements.

Specific areas to consider when developing response protocols:

- Should initial responses depend on how the potential incident is reported? For example, do we respond in the same way if a customer phones in that the water tastes funny or if we find empty drums marked cyanide next to an open hatch at a small tank? Should the protocols be detailed enough to consider different types of incident reporting?
- What pro-active responses are appropriate between the time that an incident is first identified and it is confirmed as a true contamination event? What is the balance between acting on an unconfirmed incident at the expense of supply disruption and waiting until the incident is confirmed? This is likely to be a politically charged situation and may involve local elected officials.

The primary components of a contamination response plan should include:

- Immediate response to contain and isolate the event
- On-site response by personnel capable of assessing the situation and making immediate decisions
- Notification, mobilization, and activation of internal and external sampling, testing, and response resources
- Activation of the utility's Emergency Operations Center (EOC)
- Decision support tools to assess the likely movement and spread of the contaminant in the distribution system (e.g., PipelineNet)

- Specific operational response actions to assess the situation and to characterize the credibility (type and extent of the contamination).
- Notifications that are required by law and regulation, such as the state, EPA, and the FBI
- Proactive public and media notification procedures
- Notification of nearby utilities, in case the incident is part of a larger terrorist strategy
- Methods of restoring to service any contaminated water system infrastructure and development of alternative or temporary water services or supplies (e.g., bottled water or treatment devices) if needed.

Note that contamination incidents, real or threatened, will probably be considered "crime scenes" by law enforcement. Therefore, access to facilities, even by utility personnel, may be restricted in order to preserve evidence. It is imperative that utilities coordinate with local police authorities to develop response procedures that consider both evidence preservation and protection of the public's water supply.

Following development of a contamination response strategy, the process should be tested and refined through validation exercises. Such exercises should involve not just the utility, but also external agencies identified in the response plan. After initial refinement, the water supply utility should continue to develop its response protocol in cooperation with other utilities and through interaction with national and state level stakeholders and funding organizations.

Discussion of Implementation Impediments

An Emergency Operations Plan (EOP) is a good basis for developing effective technical and contamination response procedures. In the absence of a well-defined EOP, a water utility should consider creating an EOP or refining the existing EOP in conjunction with the development of technical and contamination response procedures.

Implementation of new procedures will require extensive training. New protocols may require upgrading telephone, computer, and other communication and monitoring equipment. Installing the capability to isolate finished water reservoirs remotely should also be considered. The costs for implementing these policies and obtaining additional equipment will be a major impediment to most utilities. However, the training and equipment for contamination response should be able to play a role in the day-to-day operations of the utility and if implemented properly could also increase the quality of water delivered and efficiency of the utilities operations.

Refer to section 2.5.3 on Incident Reporting, Notification, and Response for incident reporting forms and reporting procedures and section 2.5.1 for public and media relations guidance.

Legal Considerations

Contamination response procedures should be coordinated with local, state, and national agencies to verify the quality of the proposed technical and contamination response procedures. Personnel perpetrating a practical joke, resulting in a false alarm, should be subject to disciplinary action.

Examples of Utility Security Practices and Other Background Materials

These documents are located in section 2.6.2 of Appendix B.

A. Exhibit 1 is the Alameda County Response Procedure for intrusion into their water distribution system.

B. Exhibit 2 is a response protocol developed by the California Department of Public Health for internal staff to deal with suspected contamination incidents at water systems.

C. Exhibit 3 was developed by Michael Baker Jr., Inc. as part of a document prepared for a California utility district. Information on contaminants, detection and response, as well as the protection strategy from this document may aid in the development of an action flowchart or policy.

This issue is further addressed in the recently released USEPA, "Response Protocol Toolbox," available on the USEPA website.

2.7 INFORMATION & COMMUNICATIONS SECURITY

2.7.1 Freedom of Information Act/Sunshine Laws

Objective

The safety of the nation's drinking water systems is a top priority throughout the country. Key to securing this critical infrastructure is the protection of sensitive information that could be used by domestic or international terrorists to disrupt or destroy these systems. All states have enacted laws addressing disclosure of public information. Some specify exemptions within the body of the law, while others have enacted "Catch All" provisions exempting all other statutes. In other cases, court orders expressly exclude information from disclosure.

The June 2002 Bioterrorism Bill acknowledges the need to protect some drinking water security information from general release, and establishes penalties applicable to a federal employee that releases drinking water utility security information submitted to the Environmental Protection Agency. Materials submitted to EPA are specifically exempted from disclosure under the Federal "Freedom of Information Act (FOIA)." However, most state and municipal governments have information disclosure laws that allow the processes of government groups to be open to public scrutiny. These laws, generally described as "Sunshine Laws," may apply to drinking water utilities since these utilities are often considered a part of local government. Even information protected from disclosure under the FOIA may be subject to these sunshine laws.

Key to the protection of sensitive information is the Department of Homeland Security proposed regulatory procedures, *Procedures for Handling Critical Infrastructure Information; Proposed Rule* (68 CFR 18523-18529, April 15, 2003). Comments accepted by June 15, 2003, and the final regulation will appear in the Federal Register at a future date. Finalization of this regulation could be a potentially important source of corollary information for utilities when considering the development of policy and procedures to address information protection and response to FOIA requests. This is an area that will continue to evolve and should be monitored by all utility owners and operators.

Summary of Elements

"State FOIA Laws: A Guide to Protecting Sensitive Water Security Information", was prepared by the Association of Metropolitan Water Agencies (AMWA). It addresses critical infrastructure protection issues raised by drinking water utilities relating to certain security information covered in public disclosure laws. The AMWA document offers information that public utilities may use to address open records or "sunshine" laws and FOIA and to change those policies where necessary for security purposes.

This is a five-part guide that includes:

- General Themes – Open Access and Freedom of Information in the 50 States
- Model Legislation (identifies several models and appropriate strategies)
- Strategies for Implementing State FOIA Models. Options suggested are:
    - Develop an appropriate test case
    - Stress awareness and education without pressing for legal results

- Develop coalitions among other similarly situated utilities.
- Virginia and Iowa: Recent FOIA-related Amendments
- Categorization of FOIA laws in all states and the District of Columbia.

Security information considered as records that is recommended for capture in a model includes, but is not limited to, the following:

- All relevant risk assessment information, including vulnerability assessment results and sources of threat data,
- Plans to prepare for, and to mitigate, terrorist activity,
- Emergency response and recovery plans,
- Security plans and procedures,
- Sampling protocols and procedures,
- Training, awareness, and other preparedness materials
- Corporate governance and audit records relating to security, and
- Continuity of operations, business continuity, and disaster recovery plans.

Readers of this report are urged to procure a copy of the AMWA document.

Discussion of Implementation Impediments

In the case of investor-owned utilities, some consumer groups may resist any attempt to prevent access to the utility's records, because of a concern about using security as a means to conceal fiscal dealings. However, access to records is a universal concern for all public and private utilities. Public utilities may also face resistance to attempt access to utility records based on taxpayer demands. It is therefore imperative that the utility industry monitor this security practice and regulatory outcomes as it evolves.

Legal Considerations

In general, all 50 states provide some legally protected right of access to public records. States have in most cases developed statutory rights of access, while in other cases the courts develop and analyze the extent of these rights. In a small number of states, a citizen's right of access to records is incorporated directly into the state's constitution, typically as part of an overarching right of privacy.

State FOIA Laws are not generally superseded or limited by Federal law. As a result, drinking water utilities will likely not be able to rely on exemptions in the Federal Freedom of Information Act, or the FOIA exemption passed by Congress, for protecting access to information at state levels. Members must also not assume that each state has developed FOIA principles similar to the Federal FOIA law, which the Congress passed in 1966. State FOIA legal traditions are, in most cases, older than the Federal FOIA and distinctive, having been developed based on state-based influences and policies.

Drinking water utilities should become familiar with the Federal FOIA law as they develop strategies in individual states. The Federal FOIA law continues to influence and inform State FOIA development. In some cases, state jurisdictions will rely on Federal FOIA legal and policy analysis to interpret State FOIA law provisions. In other cases the State FOIA law is patterned after the Federal FOIA law, but has expanded based more on State and local policies and interest group intervention. Finally, certain State FOIA provisions are not modeled after the Federal FOIA, but overlap in significant ways.

Examples of Utility Security Practices and Other Background Materials

None.

2.7.2 Secure Communications and Cyber Security

Objective

Secure communications and cyber security are essential to protection of utility information protection and information exchange. This security procedure includes: mainframe systems and other peripherals, file servers and applications, networks, Supervisory Control and Data Acquisition (SCADA) servers, real-time secure communications, protected information, hard-copy paper, email exchanges, computer systems, radio use, PDAs, mail and telephone communications. When contemplating secure communication, the basic intent should be to deter and prevent intrusions and monitoring of information at the utility. Should an event occur, the intent should be geared toward restoration of the disaster site, and migration to the restored site. Standard operating procedures and utility policies must be developed, established and communicated to all utility personnel in support of secure communications. These policies and procedures can have significant impacts in improving overall water facility security and ensuring that security related information does not fall into the wrong hands.

The importance of cyber-security protection for utility infrastructure and for public's health and welfare has grown significantly in the past several years. A single, knowledgeable terrorist or malicious hacker can quickly disrupt service. Public trust and confidence rest on a water utility's ability to manage infrastructure disruptions. The majority of the successful attacks on operating systems come from only a few software vulnerabilities. This can be attributed to attackers who are opportunistic, take the easiest and most convenient route, and exploit the best known flaws with effective and widely available attack tools. They rely on organizations not to fix known problems. They scan the Internet looking for vulnerable systems and attack indiscriminately,

Summary of Elements

Several issues must first be addressed to develop and establish utility policies and procedures in support of secure communications. The utility should review existing communication systems to determine vulnerabilities for the following communication tools, and determine what is the most appropriate course of action for the specific utility.

- Critical networks and infrastructure (i.e., mainframe, Novell, Lotus Notes, PeopleSoft, SCADA, all servers, telecommunication lines, Oracle Databases, Smallworld, CAD, etc.),
- Supported work processes (i.e., email, customer e-portals, processing invoices, construction upgrades, etc.),

- Mail and telecommunication systems.

Based upon the vulnerability assessment findings, policies and procedures must be developed by the utility and communicated to all employees as well as contractors and vendors at the utility. Possible threats include: malicious attack by any unauthorized access that would compromise the confidentiality, integrity and availability of data stored on the network; virus attack as asynchronous attack granting unauthorized access that compromises the confidentiality, integrity and availability of data stored on the network (often these are self-propagating such as worms attacking various content protocols); mail communications which may be affected by extreme weather conditions and/or chemicals illegally sent through the mail that cause serious contamination and delay in mail delivery; and telephone communications.

Disaster Recovery Plans and programs should be developed for business recovery and business resumption activities. These plans and programs address disaster assessment, which ascertains damage and impacts; alternative site support, which identifies an other site to support activities in the interim; restoration of disaster site designed to stabilize and rebuild system activities; and migration to the restored site, which facilitates the move back to a restored site to resume normal operations. These elements support standard operating procedures throughout the industry. The Disaster Recovery Institute (www.DRI) provides a wealth of information on the subject.

The standard operating procedure for all secure communications should be to contain the system(s) where the attack has occurred, determine the impacts to critical business, enact countermeasures, and notify appropriate personnel (employees, law enforcement, and the pubic) of the compromise and the impacts of the event.

Posted notices illustrating secure communications have been found to be useful at a number of industrial sites such as Sandia National Laboratories. It should be noted that all key stakeholders (i.e., employees, contractors and vendors) should be included in vulnerability assessment information gathering activities. For example, questions should be asked about common information exchange tools (i.e., cell phones and PDAs, etc.), sharing of passwords and securing workstations. This information is critical to knowing the utilities vulnerabilities, no matter how innocent the situation.

It is essential that firewalls be created to impede any attempted intrusions at the server and network, to protect all information including but not limited to email transmissions. Servers should be monitored to determine any attempted or actualized intrusions and reported immediately. A reporting system for such breaches should be developed, and, roles and responsibilities should be assigned to individuals with responsibility for response actions and strategies. These individuals should be trained in their assigned roles and responsibilities.

Hard-copy paper security systems should be developed, established, standard operating procedures and policies adopted and communicated to utility personnel. Hard-copy documents should be shredded and disposed of, and documentation should be maintained. Record keeping policies and standard operating secure communication procedures are critical to the success of a security program. Radios are not secure and radio frequencies are likely to be monitored by individuals outside the utility. Therefore, no real-time and/or security information should be discussed on radio. Cell telephones are not secure, while landlines are more secure. Cell telephones and PDAs should not be used to transmit sensitive information of any kind and at some industrial facilities are banned altogether. Leaving PDAs and cell telephones "on" allows individuals outside the utility to monitor utility activity.

Mail communications disruptions are a rare occurrence. Instances have delayed mail due to extenuating circumstances such as extreme weather conditions and more recently, chemicals illegally sent through the mail that cause serious contamination and delay mail delivery. Established standard operating procedures for the safe handling of mail, and identification of alternate handling site(s) will provide utilities with a more secure approach to mail communications.

Landlines can be monitored to provide an accurate record of calls. Examples are customer service representatives, hotlines or Help Desks, dispatchers and field crews. These transmissions are often record to ensure integrity and accuracy of response. It is essential that these landlines also be protected from intrusion, remain secure, and that back-up is supplied should it be required. With growth throughout the United States, area construction can result in telephone lines being cut, a pole hit or a main switch destroyed. For these reasons, a disaster recovery plan is suggested.

Utilization of record keeping activities to record and report on secure communication vulnerabilities and security solutions is essential. The utility should develop a secure means for reporting communication security incidents and approved secure communication systems established, maintained and monitored by utility security personnel.

Supervisory Control and Data Acquisition (SCADA) networks perform key functions, but also present a security risk. SCADA networks were initially designed to maximize functionality, but with little attention paid to security. As a result, performance, reliability, flexibility and safety of distributed control/SCADA systems are robust, while the security of these systems is often weak.

The National Infrastructure Protection Center (NIPC) recommends the following simple steps to enhance security:

- Create "strong" passwords and secure them.
- Make regular backups of critical data; keep copy offsite.
- Use virus protection software.
- Use a firewall as a gatekeeper between your computer and the Internet.
- Do not open e-mail attachments from strangers
- Regularly download and install security patches from your software vendors.

Several organizations are currently working on development of a prototype unit to help provide secure communications for SCADA systems. Among the organizations are the American Gas Association in concert with the Department of Homeland Security, AwwaRF, and Gas Technology Institute (GTI) in association with the National Institute of Standards and Technology (NIST).

Discussion of Implementation Impediments

Personal use of cell phones and PDAs can present issues for personnel at the utility. Keeping in touch with family members via cell phones is the most common issue. The use of PDAs can create issues dealing with secure information. It is therefore important to hear from all key stakeholders (i.e., employees, security personnel, etc.) prior to defining acceptable secure communication devices that are allowed and not allowed at the utility.

Legal Considerations

Federal, State and local laws and regulations apply to recording telephone conversations. Should the utility determine that monitoring and recording employee telephone conversations as an aspect of secure communications is necessary, this decision should be elevated to legal counsel for applicability to Federal, State and local laws. An example of a recorded telephone call would be a Help Desk or customer response hotline.

The law of intellectual property and proprietary information is still being developed for the World Wide Web (Internet). Many uncertainties exist at this time to the use of the Internet for information dissemination.

Examples of Utility Security Practices and Other Background Materials

These documents are located in section 2.7.4 of Appendix B.

A. Exhibit 1 is a booklet *21 Steps to Improve Cyber Security of SCADA Networks* from the Sandia National Laboratories.

B. Exhibit 2 is NIPC Simple Security Tips, focusing on password security.

2.7.3 Protection of Information

Objective

The objective of this section is to establish procedures that will allow access to those that have legitimate needs to the information and denying access to individuals or groups that do not have a legitimate need. This security procedure will focus on concepts and approaches to protecting information.

Until September 11, 2001, the drinking water community has generally been very open regarding work on technical issues. Security information now requires consideration of whether the information should be released. Most drinking water utility operational and engineering information are available to those with malevolent intent. While much of this information has been removed from Web sites, copies of this information could have been archived or otherwise preserved. Furthermore, operational and engineering information has been distributed by utilities as a part of ongoing and past projects, with no consideration previously given to the control of this information. This information is often still available in engineering and other offices with no special protection.

Summary of Elements

Concepts and approaches should include a determination by the utility for specific release of information, establishment of specific procedures for receipt of requests, a review of the requests, and final disposition of the submitted request. This decision making criteria should include, consideration given to the sensitivity and type of information requested, how this information could be 'used,' determination of who may have access to sensitive security information, how to mark sensitive documents, how to store and transmit sensitive security information, how to edit or summarize sensitive documents to "sanitize" documents for more general release, and, if necessary, how to deny specific information requests and justification for denials. A document handling request and disposition documentation should be developed to track these requests. This 'systematic' approach will provide a record of events to track all requests and information exchanges.

The City of Phoenix, Arizona Water Services Department has implemented an electronic procedure that allows for confidentiality of plans and specifications of city public work facilities, including water treatment plants and distribution systems. All quarter-section maps, plans, specifications and as-builts must be authorized prior to release. Authorization deems that the person receiving the plans shall not jeopardize the safety of the public or create a risk to the property. Authorization is issued by either the city's project manager or plan reviewer.

In existence is a security information protection procedural document that establishes procedures for review and protection of security information. By way of example, this procedure addresses three levels of security information:

- Very Sensitive Security Information: This is information that could be specifically useful in a malevolent attack on a specific water utility, or information that would be useful in malevolent attacks against water utilities generally. Eight different types of information are identified as "Very Sensitive".
- Sensitive Security Information: This is information that, when coupled with other information, could be useful in malevolent attacks against water utilities. Seventeen different types of information are considered "Sensitive".
- Non-Sensitive Security Information: This is security information that is not useful for malevolent attacks against water utilities. This information is either so general, or so non-specific, that it cannot be harmful in any manner. Three types of information are considered non-sensitive.

While this level of information protection may be more than is warranted for most utilities, it can serve as a guide to define terms of release and classification of documents in support of information protection standard operating procedures.

These procedures can apply to utility personnel, any project or research participant, advisory members, board members, the community at-large, and other volunteers. These procedures can also, by extension as a requirement for obtaining security information, apply to recipients of security information such as subscribers and other partners in protection of drinking water (consulting engineers, drinking water regulatory personnel, etc.). While this procedure applies to the generation, handling, and dissemination of utility information, these procedures have no authority or standing beyond an established standard operating procedure for information security handling. This concept may be applied to any utility, project, or activity that is *primarily* driven by drinking water utility security considerations.

A major investor-owned utility developed an Information Security Policies and Practices Data Classification Policy. This policy provides direction for users on how information should be classified in order to determine how to record, store, process, transmit, and otherwise handle information. This standard operating procedure may also be applied to utilities in terms of developing information SOPs. Major headings addressed in this policy include:

- Scope
- Employee Responsibilities
- Major Benefits and Risks
- Applicable Information
- Consistent Protection
- Access Control
- Classification Labels
- Labeling
- Third Party Interactions
- Shipping and Handling
- Declassification and Handling
- Destruction and Disposal
- Physical Security
- Special Consideration for Secret Information
- Violations

Finally, the Massachusetts Water Resources Authority (MWRA) developed an Information Security Policy and guidelines for controlling access to and protecting their security-restricted information from unwanted access and tampering. Their keys to the successful implementation of an effective information security strategy include:

- An established and enforceable information security policy that addresses access, storage, and sharing of information determined to be security-restricted.
- A systematic approach of classifying information considered security-restricted and a periodic review to verify the need of continued protection.
- Processes to track and monitor intrusion attempts and respond appropriately.

This policy identifies procedures for staff responsibilities; restricted information and classifications; security clearances; server security; managing security restricted documents; security for phone lines, airwaves, and verbal communications; and security for computer systems including laptop computers and traveling with laptops.

This element of information protection should include insights gleaned from Freedom of Information Act and Sunshine Laws considerations (refer to section 2.7.1 of this document), the June 2002 Bioterrorism Bill, and the recently proposed Department of Homeland Security regulatory procedures. These rules and regulations provide a framework and examples that may be applied to individual utility information protection needs.

Discussion of Implementation Impediments

Many individuals and companies have legitimate needs for some sensitive information from utilities. These include developers, prospective bidders on construction projects, plumbers, and others. The utility will have to balance legitimate needs with the protection of sensitive information.

Legal Considerations

The President's Commission on Critical Infrastructure Protection issued a report in 1997 on *Legal Impediments to Information Sharing, A "Legal Foundations" Study,* Report 10 of 12 (www.ciao.gov/resource/pccip/lf10.pdf) that identified and categorized many of the same issues now facing the water utility industry. For instance, the need for centralized guidelines, guidelines for handling classified information, guidelines for addressing trade secrets and proprietary corporate information, confidential information mechanisms, and mechanisms for addressing state "government in the sunshine" laws were identified as high priority needs. However, no follow-up took place to address these legal impediments at the Federal level, especially as would be pertinent to the drinking water community. Therefore, individual utilities will have to consult their own legal counsel to determine what restrictions may be placed on information dissemination.

Examples of Utility Security Practices and Other Background Materials

These documents are located in section 2.7.3 of Appendix B.

A. Exhibit 1 is a Records Request Protocol from the City of Phoenix Water Services Department.

B. Exhibit 2 is a Records Request Form from Phoenix. Key information includes requestor name and company; specific documents needed and the reason for request; department personnel granting request and comment box.

C. Exhibit 3 is the Information Security Policy & Practices from a major investor-owned utility. This document identifies internal policies regarding the classification of information. This can be used as a guideline for recording, storing, processing, transmitting and handling information.

D. Exhibit 4 is the Information Security Policy from the Massachusetts Water Resources Authority. Information in the report appendices included recommended "Job Aids" that are designed to assist their staff with security restricted spreadsheet and word processing documents.

CHAPTER 3: CONCLUSIONS

3.1 SECURITY PRACTICES SURVEY – SUMMARY OF RESULTS

To evaluate the current state of water utility security policies and procedures, a security best practices questionnaire was designed. As a result of detailed analysis of the data provided from survey respondents, Baker draws the following conclusions:

- Water utilities responding to the survey displayed significant strengths in the following areas of security planning and implementation:

  - Visitor/contractor access control;
  - Public tour policies;
  - Threat levels and associated response;
  - Chemical delivery and testing procedures;
  - Coordination with external agencies such as law enforcement;
  - Response plans threatened or actual contamination events; and
  - Cyber security plans and policies.

- The data provided by respondents suggests that there are significant weaknesses involving the following security policies and procedures:

  - Employee background checks practices;
  - Employee security training;
  - Employee safety procedures in the event of a security incident;
  - Package/mail delivery screening;
  - Coordination with the health and medical community and pharmacies;
  - Neighborhood watch programs; and
  - Information and communications security.

- There is strong evidence from this sample population that security preparation for both emergency procedures and security plans is a strong industry focal point and improving throughout the water utilities that responded to the survey.
- Written plans and procedures need to be developed to compliment security practices that have been successfully implemented within a water utility.
- Water utilities will need training and other performance enhancement assistance mechanisms in order to develop security plans and procedures that meet the criteria for having a complete plan including all critical elements.

3.2 RECOMMENDATIONS

Protecting the water industry will be important for the foreseeable future. The water industry represents a critical aspect of infrastructure protection. The mandate is to prevent acts of terrorism, reduce vulnerabilities, and maximize the ability to recover from incidents of terrorism. With sound security programs developed, established and communicated, we enhance the security of America's water supply. A vast inventory of tools exists such as the WaterISAC,

AWWA Web site, U.S. EPA guidance, Department of Homeland Security, and other associations' information, which yield constructive partnerships within the industry to combat against and provide the ability to recover from incidents of terrorism.

Most utilities perceive that a strong culture of security is essential throughout the water community, and instilling this culture ensures that new security policies and procedures are effectively implemented. Vulnerability assessments, revised emergency response plans, and installing and maintaining security improvements are critical components of this culture.

New analytical tools, research and financial assistance are needed to assist the industry in their efforts to safeguard America's water supply. This document provides guidance on what is currently available to the industry. Additional research is required. Almost daily new security technologies are researched, tested and employed which will yield critical information that must be made available to the water industry. Among the security practices, several did not yield the significant data. Those practices require additional research, and the information must be shared throughout the water community. They include:

- Employee background checks practices
- Employee security training
- Employee safety procedures in the event of a security incident
- Package/mail delivery screening
- Coordination with the health and medical community and pharmacies
- Neighborhood watch programs
- Information and communications security

The security challenges utilities face includes but is not limited to: Funding, Training, Research, Information Sharing, and Security Vigilance. These issues must be addressed and where guidance exists, plans and procedures must be established. Security is everyone's job at the utility, and accepting this culture is one of the most effective and least expensive methods we can employ to improve the security of our nations water supply.

APPENDIX A: WEBSITE SEARCHES

Security Sites

- www.infolinkscreening.com (Infolink Screening Services, Inc.)
- www.shrm.org (Society for Human Resource Management)
- www.workindex.com (Human Resources Executive Magazine)
- www.americandatabank.com (American Databank Corp.)
- www.absolutebackgrounds.com (Absolute Backgrounds, Inc.)
- www.usis.com (U.S. Investigation Services, Inc.)
- www.bna.com (Bureau of National Affairs)
- www.tisp.org (The Infrastructure Security Partnership)
- www.sans.org (The SANS Institute)
- www.watersecurity.org (Haested Methods)

Water and Utility Sites

- www.awwa.org (American Water Works Association)
- www.awra.org (American Water Resources Association)
- www.pennnet.com (Includes complete archives of *Water World* and *Water and Wastewater International*)
- www.amwa.net (Association of Metropolitan Water Agencies)
- www.acwanet.com (Association of California Water Agencies)
- www.amsa-cleanwater.org (Association of Metropolitan Sewerage Agencies)
- www.nrwa.org (National Rural Water Association)
- www.wef.org (Water Environment Federation)
- www.apwa.net (American Public Works Association)
- www.nawc.org (National Association of Water Companies)
- www.asdwa.org (Association of State Drinking Water Administrators)
- www.drinkingh2o.com (Technology Management and Information Corp.)
- www.uswaternews.com (U.S. Water News Magazine)
- www.utilityconnection.com (The Utility Connection)
- www.watertechonline.com (Water Tech Online Magazine)
- www.techstreet.com (Techstreet)
- www.waterindustry.org (Water Industry News)

Government and Government Organization Sites

- www.nga.org (National Governors Association)
- www.usmayors.org (The U.S. Conference of Mayors)
- www.governing.com (Governing Magazine)
- www.nipc.gov (National Infrastructure Protection Center)
- www.ciao.gov (Critical Infrastructure Assurance Office)

- www.epa.gov (U.S. Environmental Protection Agency)
- www.dhs.gov (Department of Homeland Security)
- www.lacity.org (City of Los Angeles)
- www.des.state.nh.us/wseb/EmergencyPlanning (State of New Hampshire)
- www.usdoj.gov/criminal/cybercrime/white_pr.htm (U.S. Department of Justice)

General Database Sites

- www.asce.org (American Society of Civil Engineers)
- www.aia.org (American Institute of Architects)
- www.ncppp.org (National Council for Public-Private Partnerships)
- www.securitybenchmarking.com (Security Benchmarking Association)
- Natural Hazards Research and Applications Center, University of Colorado
- EBSCO Publishing Co. - This is an online database for libraries that includes archives of 6,750 periodicals and 240 newspapers. Among the periodicals in the database are:

    - Civil Engineering (ASCE)
    - Engineering Management Journal
    - Engineering News Record
    - Chemical Engineering Journal
    - Water Engineering and Management
    - Ground Water
    - International Security Review
    - Information Systems Security
    - Professional Engineering
    - Risk
    - Pipeline and Gas Industry
    - Traffic World
    - Utility Week
    - Security Management
    - Security Distribution and Marketing
    - Security
    - Journal of Computer Security
    - Occupational Health and Safety
    - American Gas
    - Electric Light and Power
    - Electrical World
    - Power

REFERENCES

Author or Organization	Title	Date
"Unidentified U.S. Water Utility Reference Document"*	AWWS Threat Notification Form	n/a*
"Unidentified U.S. Water Utility Reference Document"*	Data Classification Policy	n/a*
"Unidentified U.S. Water Utility Reference Document"*	Disclosure and Authorization Form	n/a*
"Unidentified U.S. Water Utility Reference Document"*	Document Security Management Process Olympic Planning Project Terrorism Response Action Coalition	2/11/2000
"Unidentified U.S. Water Utility Reference Document"*	Emergency Communication and Media Management Plan	10/3/2001
"Unidentified U.S. Water Utility Reference Document"*	Emergency Response Manual for Public Utilities Administration Building	n/a*
"Unidentified U.S. Water Utility Reference Document"*	Emergency Response Procedures	1/7/2002
"Unidentified U.S. Water Utility Reference Document"*	Emergency Response Procedures - Biological Hazard or Chemical Release	1/7/2002
"Unidentified U.S. Water Utility Reference Document"*	Emergency Response Procedures - Distribution and water supply system	1/7/2002
"Unidentified U.S. Water Utility Reference Document"*	Emergency Response Procedures - Possible Exposure to Anthrax	1/7/2002
"Unidentified U.S. Water Utility Reference Document"*	Inspector's Guide- Physical Protection System (PPS) System Effectiveness (PE)	n/a*
"Unidentified U.S. Water Utility Reference Document"*	Media Inquiry Guide	8/26/2001
"Unidentified U.S. Water Utility Reference Document"*	Media Management/Crisis Communications Plan	n/a*
"Unidentified U.S. Water Utility Reference Document"*	Notification Procedures for Threats	n/a*
"Unidentified U.S. Water Utility Reference Document"*	Personnel Surety Memo	n/a*
"Unidentified U.S. Water Utility Reference Document"*	Policy Draft- Employment Policy	n/a**
"Unidentified U.S. Water Utility Reference Document"*	Potential Media Inquiries and Responses	8/26/2001
"Unidentified U.S. Water Utility Reference Document"*	Threat Notification Form	n/a*
"Unidentified U.S. Water Utility Reference Document"*	Water Contamination Threat (Distribution System)	1/8/2002
107th Congress	Public Health Security and Bioterrorism Preparedness and Response Act of 2002	6/12/2002
116 STAT.682 Public Law	Title IV - Drinking Water Security and Safety	6/12/2002
Alameda County Water District (ACWD)	ACWD Completes Vulnerability Assessment	n/a**
ACWD	ACWD Headquarters HVAC Upgrade (Const. Specs)	9/10/2002
ACWD	ACWD Responds to Increased Threat Level - Press Release	n/a**
ACWD	Employee Identification Card/Utility Emergency Pass	n/a**
ACWD	Mission San Jose Treatment Plant Upgrade (Const. Specs.)	11/20/2002
ACWD	Neighborhood Water Watch Program	6/17/2003
ACWD	Office Assistant I/II (Job announcement)	n/a**
ACWD	Operations Department Standard Operating Procedures	n/a**

* This came from an organization or utility that did now want to be identified
** No date was included with this document.

Author or Organization	Title	Date
ACWD	Safety and Security Procedures - Tours of ACWD Facilities	n/a**
ACWD	Sample - Background security Checks Memo	n/a**
ACWD	Security Response Actions	2/8/2003
American Chemistry Council	Protecting A Nation - Homeland Defense and the Business of Chemistry	4/1/2002
Aqueduct Magazine, LADWP	Risk Assessment	n/a**
American Society of Civil Engineers (ASCE)	The Infrastructure Security Partnership	n/a**
ASCE/The Infrastructure Security Partnership	Water Information Sharing and Analysis Center (ISAC) Launched	n/a**
American Water Works Association (AWWA)	AWWA 2002 Annual Conference- Counter Terrorism and Security in the Water Industry (Workshop Presentations)	2002
AWWA	Security Analysis and Response for Water Utilities	2001
AWWA	The Water Utility & Community Response	n/a**
AWWA	Vulnerability Assessment of Water Utilities and Protective Measures	2002
AWWA E-Journal	Security	July 2002
AWWA Journal	Creating the Denver Water Department's Security Program	n/a**
AWWA, Opflow	Water System Security - An Olympic Effort in Salt Lake City	2/1/2002
Awwa Research Foundation (AwwaRF)	Risk Assessment Methodology for Water Utilities	n/a**
AwwaRF	Security Information Protection Procedures Draft	8/19/2002
Babcock, HR Magazine	Checks and Balances	3/1/2003
Baker Document	Contamination Monitoring and Response	n/a**
Baker Document	Negligent Hiring and the need for Employment Background Checks	n/a**
Barry J. Nadell, Infolink Screening Services	Employee Files - Digging Deeper	May-01
Barry J. Nadell, Infolink Screening Services	How to avoid the Employee from Hell - Legally!	n/a**
Bay Area Security Information Collaborative (BASIC)	Chapter Article Regarding Formation of the Bay Area Security Information Collaborative (BASIC)	Aug-02
BASIC	Water Utility Threat Condition ERP for the Homeland Security Advisory	9/25/2002
Begley, Sharon. Newsweek	Make Safeguards Go With the Flow	11/5/2001
Belleview News-Democrat	Belleville, IL-Based Water Company Cites Security for Rate Hike	12/16/2003
BexarMet	Citizens Looking After their Waterworks	1/30/2003
Booth, Ron. CH2M Hill	Technical Security and Countermeasures White Paper for Water Utilities	12/12/2001
Brian Jordan and Rob Beste, AWWA Journal	Beverly Hills Tackles Smaller System Security Issues	Apr-03
California Dept. of Health services	Public Water System Inspection Checklist for Security	n/a**
California Dept. of Health Services	Response Protocol for Intrusion into Public Water Supply Systems	n/a**
Chritton, Michael R. The Military Engineer	Assessing the Security of Drinking Water	Feb-03
City of Phoenix	Ordinance G4396	3/1/2002
City of Phoenix Water Service Department	Badge and Background Check Data Form	n/a**
City of Phoenix Water Service Department	Bomb Threat Call Checklist	n/a**
City of Phoenix Water Service Department	Bomb Threat Response Procedure	9/18/2002
City of Phoenix Water Service Department	Contractor and Consultant Badging	n/a**
City of Phoenix Water Service Department	Mail Handling Procedures	2/18/2003
City of Phoenix Water Service Department	Procedures for Reporting Suspicious Incidents	Aug-02
City of Phoenix Water Service Department	Security Guard Protocol	4/2/2002
City of Phoenix Water Service Department	Threat Condition Green - Low Condition	4/17/2003

* This came from an organization or utility that did now want to be identified

** No date was included with this document.

Author or Organization	Title	Date
City of Phoenix Water Service Department	Water Services Department Management Services Division	3/27/2002
City of Phoenix Water Service Department	Water Services Department PMO Security Access Form	n/a**
City of Phoenix Water Services Department	Records Request Form	n/a**
City of Phoenix Water Services Department	Safety and Security	n/a**
City of Phoenix Water Services Department	Specifications - Section 01561 "SECURITY"	2/28/2002
Clark and Deininger, AWWA	Minimizing the Vulnerability of Water Supplies to Natural and Terrorist Threats	2001
Davis, Kathleen	Former Military Men Reveal Loopholes in Utility Security	5/1/2002
Department of the Army	FM 3-19.30: Physical Security	1/8/2001
Department of Defense	DOD Minimum Antiterrorism Standards for Buildings	7/31/2002
Department of Energy (DOE)	21 Steps to Improve Cyber Security of SCADA Networks	n/a**
DOE, Office of Security and Emergency Operations	Integrated Safeguards and Security Management (ISSM) Policy	5/8/2001
East Bay Municipal Utility District (EBMUD)	Security Policy	1/28/2003
Edmund Security and Emergency Operations Teams	Security Plan	Nov-01
Engelbrecht, Col. Joseph A. Jr., PhD	Alternate Futures for 2025: Security Planning to Avoid Surprise	Apr-96
Environmental Protection Agency (EPA)	Guidance for Water Utility Response, Recovery and Remediation actions for Man-Made and/or Technological Emergencies	Apr-02
EPA	Model Contaminant Matrix	2002
EPA	Strategic Plan for Homeland Security	Sept-02
EPA	Strategy 1: Physical and Operational Security (Part 3)	1989
EPA	Water Security Strategy for Systems Serving Populations Less than 100,000/15 MGD or Less	7/9/2002
EPA, Office of Water	Instructions to Assist Community Water Systems in Complying with the Public Health	Jan-03
EPA, New England Region	Drinking Water and Emergency Preparedness	n/a**
General Accounting Office	Homeland Security: Voluntary Initiatives Are Under Way at Chemical Facilities	Mar-03
Governing Magazine	Network Security Digital Nightmare	April-02
Hoffbuhr, Jack W. WaterScape, AWWA Journal	The Three C's	Jul-02
HR Magazine, Merry Mayer	Background Checks in Focus	1/1/2002
Human Capital Magazine, Profiles in Excellence	Taking Chance out of Hiring Process	2002
Information Security, EPA	Protocol to Secure Vulnerability Assessments Submitted by Community Water Systems to EPA	11/30/2002
Journal of Infrastructure Systems	Reducing Vulnerability of Water Supply Systems to Attack	Dec-98
Landers, Jay, Civil Engineering	Safeguarding Water Utilities	Jun-02
Maine Rural Water Association	Model Y2K Contingency Plan for Water Systems	n/a**
Massachusetts Water Resources Authority (MWRA)	Delivery Acceptance of Gaseous Chlorine	11/5/2001
Mills, Donald A. – AWWA Journal	As We See it - Rising to the Challenge	11/1/2001
Massachusetts Water Resources Authority (MWRA)	Delivery Acceptance of Gasoline and Diesel	5/22/2001
MWRA	Delivery Acceptance of Sodium Hypochlorite	11/5/2001
MWRA	DITP Tour Protocol	n/a**
MWRA	Identification and Control Methods Regarding Suspicious Mail, Parcels and Small Packages	11/25/2002
MWRA	Information Security Policy	n/a**
MWRA	Photo Identification Badge Procedure	4/18/2003

* This came from an organization or utility that did now want to be identified
** No date was included with this document.

Author or Organization	Title	Date
MWRA	Security Guidelines for the Inspection of Bulk Chemical Deliveries	10/18/2001
MWRA	Updating Card Access Procedure	4/18/2003
MWRA	Water Treatment Chemical Purchase Specifications	3/13/2003
MWRA Incident Response Guide	Draft Policy on Emergency Incident Management	3/1/2003
National Environmental Training Center	Protecting Your Community's Assets: A Guide for Small Wastewater Systems	Nov-02
National Infrastructure Protection Center	Seven Simple Computer Security Tips for Small Businesses and Home Computer Users	n/a**
National Nuclear Security Administration	Marking Classified Documents	3/1/2001
New Hampshire Department of Environmental Services	Public Water System Emergency Plan Guide - for community systems serving 501 or more people	n/a**
NGA	Top 10 Suggestions from State Health Officials Who've Been There	n/a**
Orlando Sentinel	Cost of Water Security Runs Deep in Central Florida	4/8/2003
Portland Bureau of Water Works	Emergency Operations Plan	Jul-99
Riordan, Ray. EBMUD	The Bay Area Security Information Collaborative (BASIC)	n/a**
Sandia National Labs	Computer Security	n/a**
Sandia National Labs	Facing Up to The Challenge	Jan-01
Sandia National Labs	Integrated Management at Sandia National Laboratories	2002
Sandia National Labs	Pocket Guide: Emergency Preparedness	Sept-2002
Sandia National Labs	Safeguards and Security Feedback and Improvement Program	n/a**
Sandia National Labs	Science and Security in the 21st Century	Apr-02
Sandia National Labs	Security Incident Management Program	5/15/2002
Sandia National Labs	Operations Security Assessment Data Collection Form	n/a**
Sandia National Labs	OPSEC Guidance - Before and After Construction	Mar-03
Sandia NL, Protective Service Directorate	Protecting Sensitive Unclassified Information Booklet (Complying with Need-to-Know Protects Your Job)	Apr-01
Santa Clara Valley Water District	Security Assessment Report	Feb-03
School of Public Health at New York Medical College	Bioterrorism: A Guide for Community Leaders and First Responders	Jun-02
Schreppel, Connie – Mohawk Valley Water Authority	Innovative Monitoring Device is Key Element of Water Utility's Security Plan	n/a**
Schreppel, Connie – Mohawk Valley Water Authority	On-Line Real Time Monitoring and Rapid Screening Methods - Peace of Mind?	n/a**
Security Magazine	What to do When Terror Alert Rises	Mar-03
Snohomish County Department of Emergency Mgmt.	Crisis/Disaster Response Handbook	Feb-03
States, Stanley; Michelle Scheuring; John Kuchta; Jessica Newberry and Leonard Casson	Utility-Based Analytical Methods to Ensure Public Water Supply Security	Apr-03
Steves, David – Register Guard	Oregon Lawmakers Review Utility Security Bill	3/4/2003
Thompson, Courtney – The Oregonian	Portland, Oregon; Considers Quick Remedy for Water Security	3/1/2002
Water World, Focus	Volunteers Participate in Utility Security	n/a**
WaterWorld, Water and Wastewater Technology	Mayor Hahn, LADWP announce increased security measures for LA's water and power system	2002
WaterWorld, Water and Wastewater Technology	Testing, Monitoring Key to Water System Security	2002
Water Environment Federation (WEF)	A Sense of Security, Parts 1 and 2	Feb-02
WEF	Best Practices in Utility Management: Innovation at Moulton Niguel Water District	2001
WEF	Maximum Security: Protect your Infrastructure with a Structured Approach for Security Assessment	Jan-02

* This came from an organization or utility that did now want to be identified
** No date was included with this document.

Author or Organization	Title	Date
WEF	Reducing Risks and Enhancing Security Through a Comprehensive Asset Management Approach	Sept-02
WEF	Safety Corner: A Sense of Security	Feb-02
WEF	The Effects of Sept 11, 2001 on the Las Vegas Area Water and Wastewater Systems	2002
WEF	Water Related Terrorism Unlikely, Experts Say	Dec-01
WEF/AWWA	Terrorism: How to Use Technology to Deter and Detect (D & D)	2002
Wettering, Larry. Opflow	Lessons Learned: Taking Security to the Next Level	Feb-02
White House	The National Strategy for the Physical Protection of Critical Infrastructures and Key Assets	Feb-03
Zeichner Research Analysis	State Implementation of Federal Cyber-Security Requirements	n/a**

* This came from an organization or utility that did now want to be identified
** No date was included with this document.

ABBREVIATIONS

ACWD	Alameda County Water District
AMWA	Association of Metropolitan Water Agencies
ASCE	American Society of Civil Engineers
AWWA	American Water Works Association
AwwaRF	Awwa Research Foundation
BASIC	Bay Area Security Information Collaborative
CALWARN	California Water/Wastewater Agency Response Network
CCTV	Closed Circuit Television
CDC	Centers for Disease Control and Prevention
CLAW	Citizens Looking After their Waterworks
CRA	Consumer Reporting Agency
CRWA	California Rural Water Association
CSMRA	California Sanitation Risk Management Authority
CUEA	California Utilities Emergency Association
DEP	Department of Environmental Protection
DHS	Department of Health Services
DOD	Department of Defense
DOE	Department of Energy
EBMUD	East Bay Municipal Utility District
ECT	Emergency Communications Team
EMS	Emergency Medical Service
EOP	Emergency Operations Plan
EPA	Environmental Protection Agency
FBI	Federal Bureau of Investigation
FCRA	Fair Credit Reporting Act
FOIA	Freedom of Information Act
GIS	Geographical Information Systems
GTI	Gas Technology Institute
HAZMAT	Hazardous Materials
HSAS	Homeland Security Advisory System
ID	Identification
IT	Information Technology
MDSD	Material Safety Data Sheets
MWRA	Massachusetts Water Resources Authority

NIPC	National Infrastructure Protection Center
NGA	National Governor's Association
NIST	National Institute of Standards and Technology
NL	National Laboratories
NRC	National Response Center
OES	California Office of Emergency Services
OSHA	Occupational Safety and Health Administration
PAC	Project Advisory Committee
PDA	Personal Digital Assistant
PIO	Public Information Officer
PPE	Personal Protective Equipment
RSC	Regional Steering Committee
SCADA	Supervisory Control And Data Acquisition
SOP	Standard Operating Procedures
SSC	State Steering Committee
US	United States
WARN	Water/Wastewater Agency Response Network
WaterISAC	Water Information Sharing and Analysis Center
WEF	Water Environment Federation

IWA Publishing's authorised EU representative for General Product Safety Regulations is Diane D'Arras, 15 rue Duret, 75116 Paris, France, e-mail: safety@iwap.co.uk.

Printed and bound by CPI Group (UK) Ltd, Croydon, CR0 4YY
17/09/2026
02218802-0020